AF572468

# BEEKEEPING
## For Fun and Profit

# BEEKEEPING
## For Fun and Profit

Louise G. Hanson

*Photographs by Mike Nielsen and Dan Westesen*
*Line Drawings by Brent Burch*

*David McKay Company, Inc.*
NEW YORK

**Library of Congress Cataloging in Publication Data**
Hanson, Louise G
Beekeeping for fun and profit.

Includes index.
SUMMARY: Basic facts about keeping and working with bees and harvesting and selling honey.
1. Bee culture. [1. Bee culture] I. Nielsen, Mike. II. Westesen, Dan. III. Burch, Brent. IV. Title.
SF523.H24 638′.1 79–23495
ISBN 0-679-20529-2

1 2 3 4 5 6 7 8 9 10

MANUFACTURED IN THE UNITED STATES OF AMERICA

# Contents

# Foreword

As a boy, I liked honey very much. My father would let me take a stick and dig it out of the big drums and cans. Becuase I liked the product so much, I guess that's what got me started in the business.

Dad was a philosophical old pioneer. He taught me the philosophy of the bees — *to save*. Bees save everything: pollen, beeswax, and honey. But they are not selfish about it. After they fill a hive, they swarm and leave everything to the next generation. Honey producers have to be conservative in this business, too; they must save and use everything in order to be successful. This idea fits with the philosophy Dad constantly taught about the bees, as he prepared a storehouse of knowledge to leave to his next generation.

Many aspects of beekeeping are fun. For example, I enjoyed improving the strain of bees, as cattlemen do with Hereford and Angus cattle. I personally bred the bees down so they wouldn't sting. Although they retained their stingers, they were beautiful stingless bees. Unfortunately, I took the pep out of them. They became so lazy they not only wouldn't sting but they also wouldn't gather honey. So I had to back up — to select and select until I got back to a more normal type of bee. Our present bee is good-natured and doesn't sting. It is called the Miller Honey Commercial Hybrid — a cross between a Caucasian and an Italian bee, with an injection of the Carneolian bee from the Carneolian Islands. So that was one phase of the business I had fun with.

*Woodrow Miller, largest honey producer in the world, at his desk in Colton, California.*

As for profit in honey production, in 1934 when our family started in the bee business, it was supposedly the worst enterprise we could go into, with the exception of raising goats. Now it is literally a gold mine; I have made a fortune out of honey production. The potential in this business is still great today.

*Woodrow Miller*
*Colton, California*

# What This Book Is About

Lots of activites are fun. If you are so inclined, skiing is fun; swimming is fun; bowling is fun. If you are otherwise inclined, chess is fun; reading is fun; sewing is fun. This book will introduce you to an activity somewhere between skiing and sewing that you might find not only fun but, perhaps, profitable.

It can be fun, for example, to learn something about the biology of a bee, about becoming partly self-sufficient, about bees and the environment, about the language of beekeeping. It can be more fun making beekeeping a hobby — taking care of your own bees and harvesting the honey they produce beyond their needs. It can even be fun for you to imagine what it might be like to go into honey production as a career.

This book begins with little-known, interesting facts about bees; continues with ideas about self-sufficiency, ecology, and beginning beekeeping; and concludes with suggestions on how to become a commercial honey producer.

If any one of these ideas appeals to you, read that part of the book first. You might decide to read all of it.

# PART I

# Fascinating Facts about Beekeeping

# 1

# Having Fun Learning about Bees

Do you know who does all the work in — and out of — a beehive? A lot of little female bees, that's who. Although feminists may object to the female bees having to do all the housework and shopping, it's one situation they can't do much about. It's nature. Besides, there is nobility and pride in hard work.

Just for the fun of it, sometime in the middle of a warm day in late spring or midsummer, take a friend to see a beekeeper (an *apiarist*) and ask if you can see the inside of a hive. If you're a little nervous, ask for a bee veil and gloves so you won't jump when a bee gets near you. The calmer you are, the calmer the bees are. And on a warm day when the wind isn't blowing, the bees will be busily coming and going and won't pay much attention to you.

Ask the apiarist to pull out a frame of bees (a thin sheet of wax, called *foundation,* framed with wood). By late spring the foundation will probably be partly — or mostly — filled with honey. Bees will be crawling all over it, as if they didn't have any sense. But they know what they're doing. These are female bees.

If the frame is from a *super* (a box of frames sitting *on top* of the brood box where the queen is laying her eggs), these female bees are called *worker* bees. These bees are busy making honey and storing pollen for winter.

*Worker bees busily storing, ripening, and covering honey with beeswax.*

If the frame is from the brood box, the busy female bees are *nurse* bees, who are feeding the queen and her brood.

The bees flying in and out of the hive are the female *field* bees. They're the ones who go out in search of nectar, pollen, and water to bring back for the nurse bees and the worker bees to use. Technically, of course, the field bees are worker bees too, but calling them field bees helps to distinguish them from the other worker bees. Usually, in less than six weeks, the wings of these field bees become tattered from flying and foraging, and they die. These little female bees can also lay eggs without mating. This process is called *parthenogenesis* (from the Greek *parthenos,* virgin, and *genesis,* origin). But they don't lay eggs very often because they don't like drones (for the reason, see the fourth paragraph from this one), and that's the only kind of egg they lay. Somehow, though, they become disoriented and confused when they suddenly lose their queen, and they begin to lay some drone eggs.

Now that you've watch the apiarist handle a frame, ask if you can hold it so you can inspect it more closely. But beware. You may catch the beekeeping fever. I did. The first symptom is an uncontrollable desire to

*Field bees bringing their stores to the hives.*

*A beginning beekeeper examines a frame of bees and honey.*

have bees of your own. If you don't get the fever under control, by the spring of the year you'll turn into a beekeeper. And the fever is usually irreversible. You stay that way the rest of your life.

The queen bee of a hive does nothing but lay eggs. She doesn't even feed herself. If the nurse bees didn't put food into her mouth, she'd starve. She's longer and bigger than all the other female bees, and she lays eggs every day after she goes on her maiden flight. As she soars into the sky, she mates with one or more drones; then she never mates again. She lays eggs from one- to five-year periods — 1,000 to 3,000 a day until she grows older. Many commercial apiarists requeen their hives every year — or every other year — for maximum hive strength.

So where does that leave the drone — the only male bee in the hive? You will see quite a few drones in one hive, though not as many as the number of females. Drones are bigger than the industrious females and thicker than the queen, though no so long. They do not have stingers; so you don't have to be afraid of them.

As I said, they aren't very popular. They hang around, getting in the way, trying to get food. Since drones are of no help to the busy little female bees, they get pushed out of the hive before winter sets in. While the mating season is on, drones get pushed to the edge of the frame. There they lurk, hoping for a handout. And when a drone does fly after a queen and mates with her, the act of mating rips out some of his organs, and he plunges to the earth, dead. His usefulness is over. Who says it's a man's world? It certainty isn't in a beehive!

There is some poetic justice here, though. When a female worker bee stings someone, her stinger pulls out a muscle from *her* body, and *she* dies, too. So the next time a bee stings you, don't be too angry at her; she has given her life in return for the act.

Here are some other facts you should know about bees.

- Female worker bees take about 21 days to hatch.
- The queen bee takes only about 15 days to hatch. This is because she is fed *royal jelly* — a secretion (something that separates) from the nurse bees that is rich in nutrients and hormones — while she is developing.
- Drones take 24 days to hatch.
- The stages of a bee's birth and life and cycle are 1) egg, 2) larva (larvae, plural), 3) pupa (pupae, plural), and 4) fully developed bee.

*The queen bee (in center) attended by nurse bees, who feed her royal jelly.*

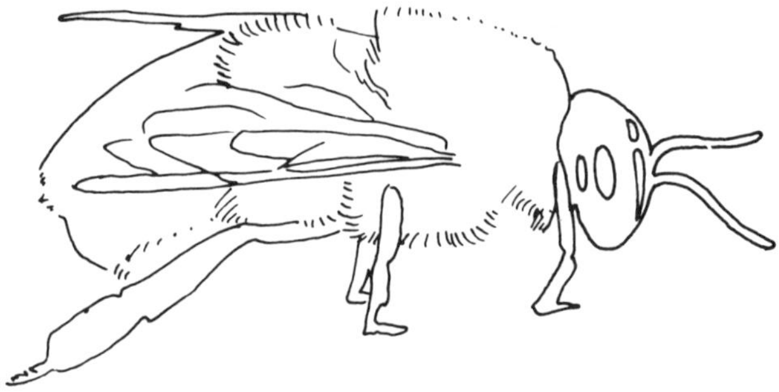

*The thickset drone, or male bee.*

*Worker bees, drones, and the queen.*

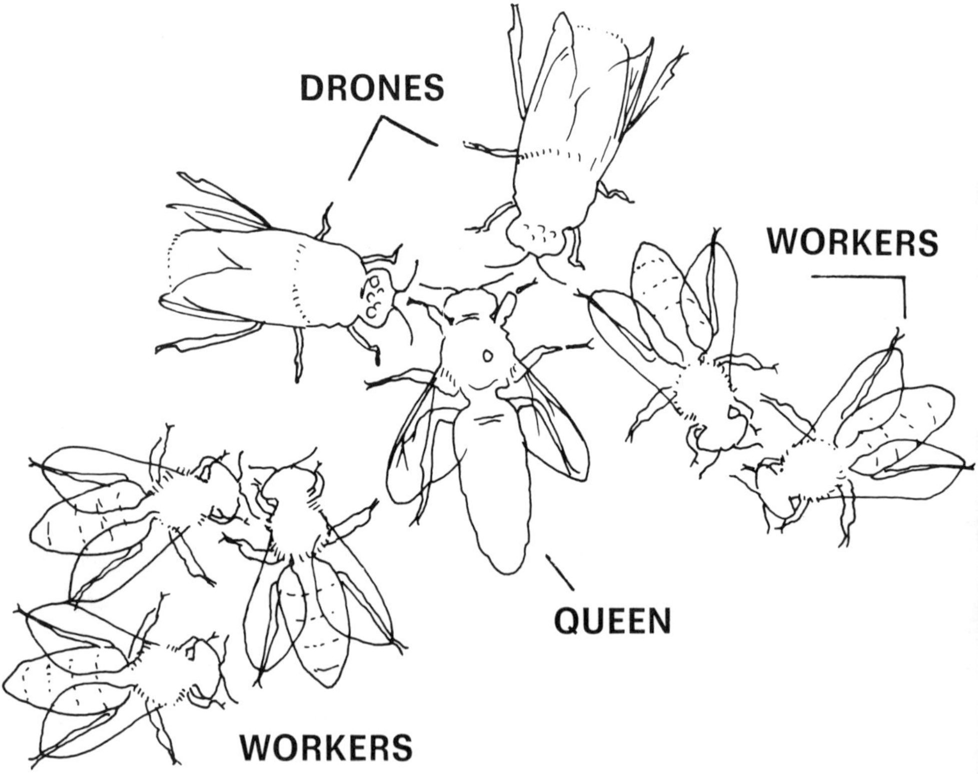

→

*Larvae (uncapped, curled in cells). The pupae are covered with beeswax, waiting to emerge as bees. A few tiny eggs may be seen with a magnifying glass. These are all called* brood.

*A field bee collecting nectar and pollen (note the pollen in the pollen basket on her leg) from a thistle blossom.*

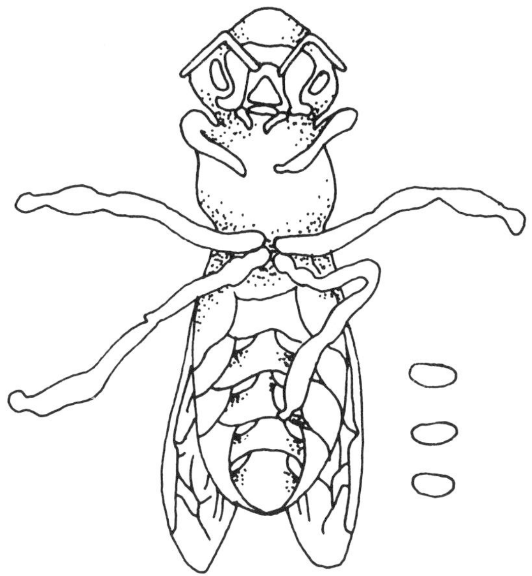

*Wax glands of the worker bee.*

- A honeybee makes 154 trips for one teaspoon of honey.
- A bee flies a distance equal to more than three times around the world to gather a pound of honey.
- A pound of ripened honey takes 160,000 bee hours.

Every female worker bee goes through the same cycle of chores as every other female worker bee.

From the first day of her life she begins to clean cells in the wax foundation for the queen to lay her eggs in. In a few days she becomes a nurse, caring for the queen and feeding the larvae. After that, she becomes a worker, storing honey. When she is 15- to 20-days old, she is entrusted with doing the shopping: becoming a field bee foraging for nectar, pollen, and water. She remains at this work for several weeks before she dies.

Beeswax is secreted from the worker bee's abdomen. She has four pairs of wax glands in her abdomen, from which she takes the almost microscopic flakes of wax with her mandibles and uses them to "draw out" (shape and make deeper) the cells in the comb, or wax foundation, to prepare them for brood or honey.

# 2

# Pride from Self-Sufficiency

As I searched for beekeepers to talk with about bees and to ask if I could take some pictures for this book, I found a place not of this decade.

My husband and I, along with my photographer, headed out a graveled road south of our town. As we drew near our objective, we kept our eyes open for two tall eucalyptus trees. That would be the Robinson's place, and a friend had told me that Bill Robinson had bees.

We found the trees, and we found Bill and his bees. But that wasn't all we found. Bill and his wife had built a self-contained little town of two people on about five acres of land. Fruit trees shaded a small, white frame house and dropped their limbs onto the roof of a barn. There were two cows for milk, chickens for eggs (and eating), and a goat tethered to a tree. In the distance we could see a corral full of pigs. A large garden flourished under a fine mist of water from a nearby well: vegetables, squash, rhubarb — a half acre of bursting health.

Bill took us down to his basement and showed us jars of fruit and vegetables he and his wife had bottled. He showed us stacks of wood to burn in his stove. And he showed us his jars of honey, extracted from the several hives of bees we had seen under trees and on the roof of his barn.

"I got tired of asphalt and brick and smog and crowded freeways," he said. "This place is balm to my soul" — this from a former aircraft engineer! We listened to the silence around us, took in the clearness of the sky, and shared his peace.

*Bill Robinson, with some bee-made comb — good only for chunk honey. Bill had left a super sitting empty, and the bees made their own comb inside the lid.*

*A jar of chunk honey that Bill Robinson gave us.*

*A health enthusiast, reading about honey as a substitute for sugar.*

*An energy-using, dead-heat foot race. Honey helps.*

Within his words lay unspoken pride in his accomplishments. "I could get snowed in for two winters and never feel a pinch," he told us. We coveted what he had found there near the two tall eucalpytus trees.

Circumstances, along with city and county zoning ordinances, prohibit some of us from making the break away from the big city as Bill and his wife had done. But if we can't become totally self-sufficient, we can find ways of becoming more self-sufficient than we are. Owning bees is one of these ways.

I almost said *manufacturing* our own honey is one of these ways. But of course, *we* don't manufacture the honey; the *bees* do. And if you've ever paid for honey (or sugar) in a grocery store, you know having your own bees will save you money.

Honey can be substituted for sugar in most instances. A very good source to consult on this subject is Lily Davis's section, "Cooking and Canning with Honey," in the book she and I did together: *The Basic Beekeeping and Honey Book* (The David McKay Company, Inc., 2 Park Ave., New York, N.Y. 10016). Reasons for using honey instead of sugar, not the least of which is good health can also be found in that book. And honey is an excellent quick energy source for active teenagers.

Right now self-sufficiency is a subject on the minds of many people. So is ecology, but we'll touch upon that in the next chapter.

# 3

# The Importance of Bees in Our Ecology

Imagine, if you can, what would happen if fruit trees, farmers' crops, and flowers were not pollinated for just one year. The scarcity of food would be disastrous and the lack of color depressing.

Granted, other insects besides bees, such as butterflies, pollinate blossoms of plants and trees. But bees are one of the most productive pollinating insects. In the United States, billions of dollars of crops depend on pollination by bees. In the city, in the country, in the mountains — wherever you go in the spring and summer — you hear the buzzing of industrious bees among the blossoms.

Have you ever noticed that when you have a personal stake in something, you fight for it much harder than if you weren't involved? Perhaps, for example, you usually aren't interested in voting in school elections. But this year you hear that one candidate for student-body president plans to do away with pep assemblies. Another candidate is *against* doing away with them. You agree with this candidate; you like the way pep assemblies fire you up to give your all for your school. Therefore, you make sure your vote is cast for the candidate in favor of pep assemblies. You even do some campaigning for that candidate.

That's the way it is when you have bees of your own. You do everything you can to keep people from using sprays on fruit trees and crops that might kill your bees. You make sure your bees have sugar water and pollen when the winter is long and the blossoms are late. You fight to protect your bees. And when you do this, you make a contribution to your community's welfare. And that's not a bad thing to do.

*A bee pollinating blossoms as it collects pollen and nectar.*

# 4

# The Language of Beekeeping

- An *apiarist* is another name for a beekeeper.
- An *apiary* is several colonies (hives) of bees.
- A *bee brush* helps an apiarist brush the bees off a comb before he or she removes the honey.
- A metal *bee escape* allows bees to leave a super but prevents them from returning to it (so the beekeeper can take the honey).
- A *movable-frame hive* helps a beekeeper remove, inspect, and extract honey from frames.
- A *foundation* is a comb made of beeswax and stamped with hexagonal cells, ready for the bees to ''draw out'' the cells (shape them deeper with the wax foundation and their own beeswax).
- A *brood box* houses the foundation where the queen bee lays her eggs. As these eggs develop, they are called *brood*.
- A *queen excluder* is a wire grid placed over a brood box through which only the small workers can pass to make honey in the box above. (The queen is too large to pass through; so she and her brood remain in the brood box.)
- A *frame* is foundation surrounded by a wooden frame that fits into a super or a brood box.

*The author, a practicing beekeeper, removes a frame from one of her hives.*

*An apiary in Provo, Utah.*

*A bee brush is only one way to remove bees from a comb.*

*The pencil is pointed to the hole into which the bees "escape" from the super of honey. The bees turn both to the right and to the left, passing through springs that let them pass one way only. The springs prevent the bees from returning to this part of the hive. The bee escape is set into the hole in the center of an inner lid.*

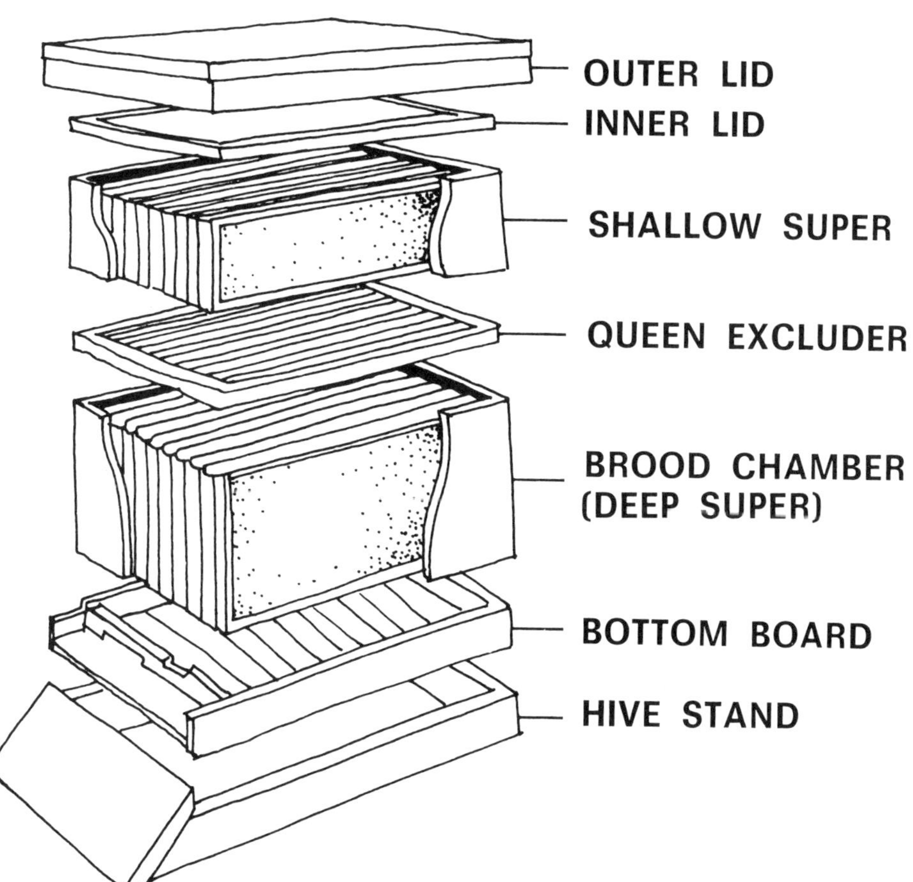


*A movable-frame hive. Evenly spaced frames of wood (holding foundation wax) are placed in the hive body and the super or supers. The frames may be used again and again.*

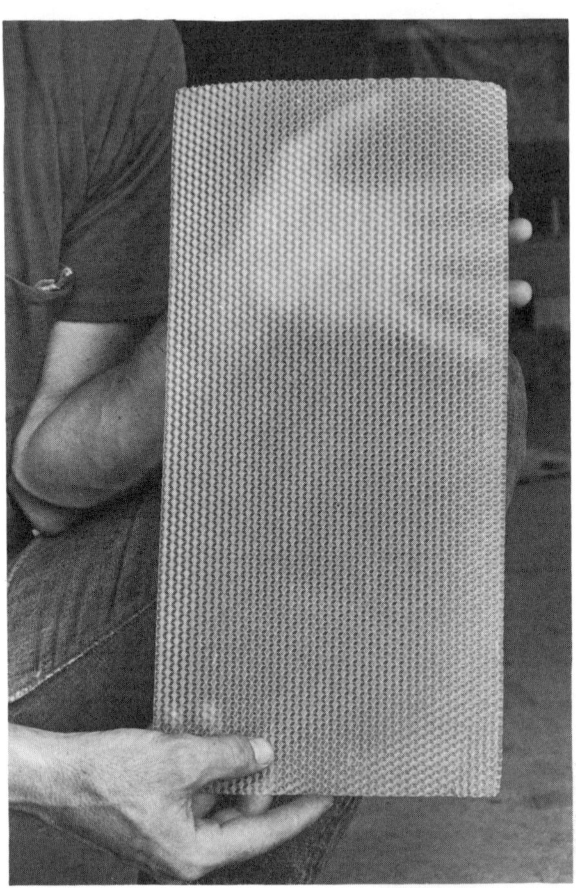

*A manufactured sheet of foundation comb.*

*A queen excluder. (Note the propolis — bee glue — placed there by worker bees.)*

*A "frame," consisting of manufactured foundation comb, reinforcing bee wire, and the wooden frame surrounding them.*

*An apiarist using his hive tool to pry apart frames that the bees have glued together with beeswax and propolis.*

*A smoker, showing how the bellows work to force the smoke from the funnel.*

- A *hive tool* is indispensable to an apiarist for prying apart frames and supers when he wants to service his bees or take out honey.
- A *smoker* is a metal fire pot with a directional funnel at the top and a bellows on the side, used for smoking bees out of the way when you inspect the hives.

The purpose of the next section of this book is to show you how much fun it is actually having your own bees.

# PART II

# The Fun of Beekeeping

# 5

# Learn Something Before You Start

Gathering some information about beekeeping before you begin can help you become a more successful beekeeper. Here are some ways to help you learn.

## Take Some Pictures

Taking snapshots of almost anything is fun. But when you visit an apiary, taking pictures can be informative as well as fascinating. One caution: if you want to take off your bee gloves to snap the pictures, try to borrow a camera with a telephoto lens. Usually, bees are busy working and will not notice you, but it's better not to take chances. For example, although he was standing a distance away, my photographer took a picture of a bee pulling itself from the cell for the first time. He might not have been able to get the shot if he had been close enough to excite the bees.

Ask the apiarist to point out what the bees are doing. He can show you how to get some interesting pictures. And you'll learn a lot of new things as you snap away.

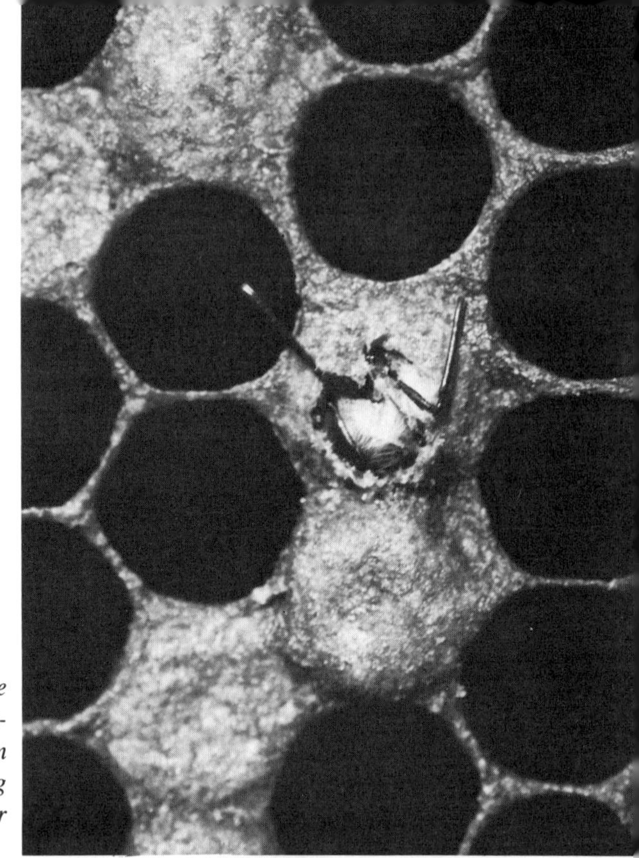

*A bee beginning to emerge from its capped cell.*

*A group of beekeeping students, one of them taking a picture of a demonstration hive. The bees have been removed, and Joel Wright is prying up a frame of honey and brood for inspection.*

# Find Out Which Nectar and Pollen Plants Grow in Your Area

If you can find a beekeeper or two near you, you can be certain nectar plants are also close by. But if you are planning to be a hobbyist rather than a commercial apiarist, you will have to know if your bees can get enough nectar for the spring and summer season without being moved to another location for a part of the season. This does not mean that hobbyists do not move their bees to better nectar fields, but some prefer to keep their bees in one place, especially if that place is their own backyard. This will be your own choice to make. To find out about nectar plants near you, write to the State Entomologist, Department of Agriculture, at the capital city of your state. If your state does not have an entomologist, someone at the department of agriculture will answer your letter and will tell you where to get the information.

If your bees do not have enough nectar and pollen, they cannot feed the brood and produce honey. In the fall, a colony of bees (one brood box with supers) should have at least 70 pounds of honey to carry them through the winter. If the summer has been too dry to produce nectar plants, or if a frost destroys the first nectar flow (an ample supply of blossoms that bees like), you will have to feed sugar water and pollen to your bees for a few weeks. Realizing how easy this is will help you decide that you can do it.

Use 1½ parts sugar to 1 part water. Heat the water just enough to dissolve the sugar. Use either a Boardman entrance feeder (an upside-down quart jar with small holes in the lid that goes into a box that fits into the entrance of the hive) or a division board feeder (a hollow plastic feeder that takes the place of a frame in the hive). The division board feeder holds more than 2 quarts of syrup. The entrance feeder holds only one. You will have to return more often to replenish the entrance feeder, but it is easier than opening the hive to refill the division board feeder. Again, you may decide which you would rather do.

You will also have to buy some pollen, either at a bee supply house near you or at a health food store. Shake 3 or 4 heaping tablespoons of pollen between the frames of the hive so that it falls onto the bottom board. The bees will mix it with honey to make "bee bread," which they store in the cells for feeding the brood. Do this at least once a week.

In these nectarless seasons you will have to leave the two supers of

*Boardman entrance feeder.*

*Division board feeder.*

Jar of pollen.

Shaking pollen into the frames.

honey for the bees to winter on. But when the nectar seasons are good, you should keep the third super (and perhaps a fourth) for yourself.

## Read a Little

- I have already mentioned that you can send to your state or federal department of agriculture for information on beekeeping.
- Another source is your state beekeepers association. Again, your state department of agriculture can let you know its address.
- Two monthly beekeeping journals with up-to-date information for beekeepers are the *American Bee Journal,* Hamilton, Illinois 62341, and *Gleanings in Bee Culture,* Medina, Ohio 44256.
- Go to the library and check out books on beekeeping.

*Reading about bees isn't boring; it's fun.*

- Visit bookstores and ask for books on beekeeping. There you will find the latest information on the subject.

When you begin to expand your knowledge about bees, you also expand your enthusiasm to begin keeping them.

## Take a Short Course in Beekeeping

Anything you do to gain more information on beekeeping will help you take care of your bees when you get them. Try one of these sources for a quickie class in the subject:

1. State agricultural colleges often have extension (home–study) courses. Write to a college in your state and find out if they have one.
2. If you can't find classes to take, get together with a few of your friends and offer a token tuition to an apiarist in your area. He may be glad to give you a few lessons in beekeeping–at his convenience, of course.

And remember: knowledge takes away fear.

# 6

# Finding a Place for Your Bees

When you're planning to have your own bees, it's natural to want them as near you as possible. For one thing, they are like new playthings: you want to be able to look at them even when you're not playing with them. For another, you want them near enough so you can service them easily. Perhaps you *can* install your bees in your own backyard or on your garage roof. So much the better if it works out that way for you.

But if you want to be sure of having healthy bees — and happy neighbors — there are some things to consider before you decide on the location of your hives (or *hive* — it may be easier to take care of only *one* hive the first year).

## For Healthy Bees

You choose a satisfactory home for your bees in the same way you choose one for yourself — a home that is pleasant and comfortable. But bees need a few things for their home that you don't, as the following sections will show.

### *Nectar and pollen*

I suspect you're saying, "Yes, I can do very well without nectar and pollen, thank you." And except for the esthetic pleasure colorful blos-

*Being able to service your bees in your own backyard is convenient.*

soms bring you really don't need them — at least not so much as your bees do. Furthermore, your esthetic sense is satisfied by almost any colorful blossom. Not so with bees. They need certain *kinds* of blossoms to make certain kinds of honey.

So what do you do about this need? A beehive should be within a 2-mile radius of the nectar the bees need. The nearer the better: it will take less working time for the bee, and the bees' wings will not wear out as quickly. Remember, you are trying to get the most honey possible from your bees. If they can make more honey than they need to get them through the winter, the surplus is all yours. That means you must leave at least two shallow supers or one deep super full of honey on top of the brood box for the bees' winter stores. If you can get the bees to fill up another shallow super — ten frames! — all that honey is yours. And what

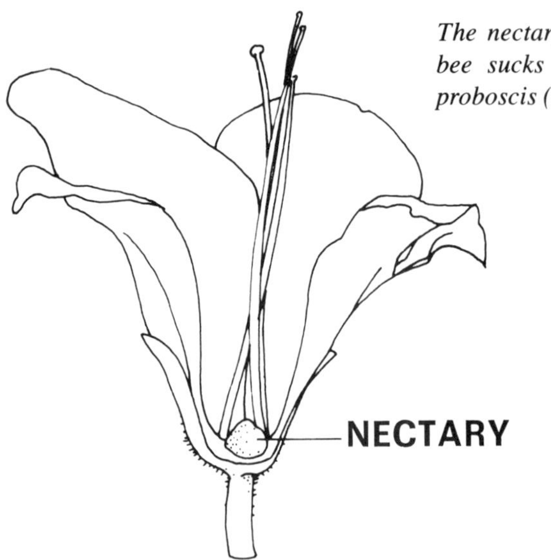


*The nectary of a flower, where the bee sucks up the nectar with its proboscis (long tongue).*

if you can get them to fill yet another super? Do you know that some beekeepers know enough to get five extra supers for themselves? Of course, that's very unusual. But that's championship beekeeping—and that's another story. (See *The Art & Adventure of Beekeeping* by Ormond and Harry Aebi, listed in the bibliography.)

Let's get back to providing the right kind of nectar for your bees.

1. As we discussed earlier, you should ask nearby apiarists what kind of nectar their bees use. Also ask if it is near enough for your bees. If not, what kind of blossoms *are* near you?
2. If you can't locate any beekeepers, write to your state's department of agriculture. I did that, and I received a booklet called *Nectar and Pollen Plants of Utah* by W. P. Nye. This booklet has pages of charts set up in this way for plants:

| Blossoming period | Region of occurrence | Usual habitat | Value as nectar source | Value as pollen source |
|---|---|---|---|---|

*A nectar plant in Utah — white clover.*

It also has a map of Utah, showing the four major beekeeping areas. With this valuable booklet I could tell when important nectar plants bloom — and where. I also found out that if I wanted a really productive yield of honey, I would have to move my bees in certain months to where the better nectar fields were.

And note that this booklet also names the best *pollen* plants. Pollen is an important source of protein for bees. Without it, the brood would not grow.

### *Water*

Bees use great amounts of water, not only for making honey but also for cooling the hives. However, make sure that it is *not* standing water. Bees defecate (expel their feces) into it, and germs multiply.

The best source of water — which should be about 100 yards from the hive to avoid contamination from flying bees — is water barely running on a board, providing constant movement to avoid stagnation. In areas of drought, a handyman could rig up a system for recycling the water.

### *Site on lot*

If you live in a cold climate, you should place your hive where it will be warmed by the winter sun and where the bees can come out to excrete their feces without dying from the cold. In the summer, the bees should have a shaded place so that they will not suffocate from the heat. If you have a spot that receives sun in the winter and shade in the summer, you have this problem solved.

Another advantage is to place your hive on a slope facing south, if it's possible. In cold weather the slope will protect the hive from north winds, and the southern exposure will allow the winter sun to warm the front of the hive. But if you don't have a slope facing south, a slope facing east will do almost as well. Build a slatted fence on the north and west sides to protect the hive from the wind. If the fence is not slatted, the wind will travel upward from the outside and downward in the inside to hit the hive.

Placing your hive on a stand also has many advantages. If you live in a snowy area, a 16-to 18-inch stand will likely keep your hive's head above the level of the snow. If you drill a small hole (½ inch) near the top of the super, the hive remains ventilated and allows the bees to go on a

*Water barely running on a board will avoid stagnation.*

*This hobbyist's hive is nestled under a tree, shaded from the heat. (Weeds should be pulled from in front of the entrance to help the bees soar in for an easy landing.)*

cleansing flight at least once a month in winter on a relatively warm day. (Bees will not defecate in their hive.) A stand also helps keep out mice and ants and discourages skunks from attacking and eating the bees. And if you gravel the area under and around the hive, you keep down weeds that otherwise would interfere with the comings and goings of the bees.

***Protection from lethal sprays***

If a major source of early spring nectar is a fruit orchard near you, you must do whatever possible to combat ill effects on your bees.

- Talk with the orchardists in your vicinity. They know they need bees to pollinate their orchards, or they will have no fruit. If they must spray, ask them to spray at night when your bees are not active. A certain kind of spray is not harmful to bees after it dries. (But if it rains, the dried spray becomes moist and lethal again — if it is not washed off completely.)
- You might do as my county bee inspector does. Knowing that fruit growers in our area hire commercial sprayers who can spray only during daytime working hours, he stuffs grass into the entrances of his hives in the morning before the spraying begins. (You wouldn't do this on a hot day, or the bees would smother. But spraying is usually done in the early, cool springtime.) By the time the spray dries, the bees will have chewed through the grass; or you can remove it for them.
- Make a friend of your county bee inspector, if you have one. He or she can let you know the kinds of sprays being used in your area and what you can do about them.

# For Happy Neighbors

Most neighbors of people with bees are unaware of the good bees do. They are, however, uncommonly knowledgeable about the *bad* they do: stinging, soiling, making a nuisance of themselves at canning time, using swimming and wading pools, bird baths, and similar sources of water that attract them. And if you have a commercial apiarist near you, he is convinced your bees will bring that dread disease, American foulbrood, to his apiary.

Some neighborhood beekeepers keep their bees so well hidden behind high fences that no one in the neighborhood is aware they are there. If neighbors find bees in their flowers and watering places, they assume the bees are simply part of nature.

If you are not so lucky, if you want to keep bees in a residential area and don't know whether you can solve the problems presented by an unhappy neighbor, you may want to consider another location for your bees. But before doing this consider the following suggestions.

- Tell your neighbors you will build an 8-foot fence around your hive and that you will face the hive away from the neighbors, toward high shrubbery where the bees will have to fly high toward their nectar sources.
- Tell them that when bees are gathering nectar or water, they are usually happy and will not sting unless they are swatted at.
- Assure any commercial apiarists that you will feed medication to your bees in the spring and the fall to avoid disease.
- Tell your neighbors bees do not usually sting when they are swarming. Show them this illustration of bees swarming over a person's arm.
- Tell them that if they are stung, they should not pinch at the stinger to remove it. Instead, they should take a sharp knife or a razor blade and slice under the sac on the surface of the skin. This will prevent the poison from draining more into the skin. The stinger may then be removed with a needle, as if it were a sliver of wood.
- Promise them you will keep a watering place for your bees, so they will not use the neighbors' watering places.
- Tell your neighbors that when they are canning, you will feed your bees sugar syrup if the honey flow is low. That will keep your bees away from their kitchen door and windows.
- Tell them you would like to share some of your honey with them. Then follow through. Take some honey to them, along with some hot bread or biscuits and butter.

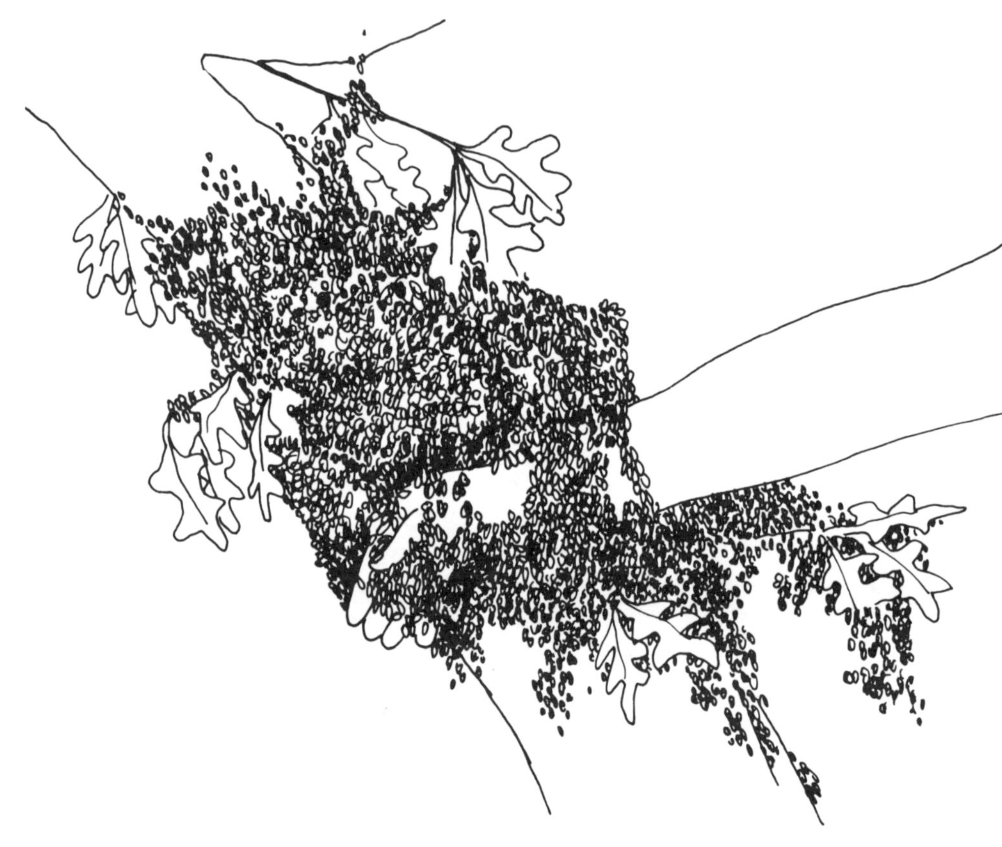

*Bees are usually gentle when they are swarming. (This drawing was made from a photograph.)*

- If the city planners have told you not to keep bees because of the neighbors, arm yourself with information from county, state, and federal departments of agriculture that proves how necessary bees are to ecology. Also, take along some letters of goodwill from your neighbors. Who knows? You may convince the city fathers that you should be allowed to keep bees.

## An Alternative Site

If you should decide not to keep bees in your backyard, try one of these ideas.

- Search out a fruit grower near you and ask if he would like to have your bees in his orchard to pollinate his fruit blossoms.
- A farmer just outside of town may like to have your bees to pollinate his crops.
- See if you can find out who owns a vacant lot in an agricultural area near you. Perhaps the owner will let you rent a small part of it for your bees.

So if you're beginning to read a little abut bees, you've visited an apiary or two, and you've decided where you would like to keep your own bees, the next chapter will tell you how to get everything ready for them.

# 7

# Readying a Hive for Your Bees

We've talked a little about the hive itself — that it should be placed on a stand, that it should be situated comfortably, and that it should be faced away from prevailing winds. This chapter will tell you more about how the hive works and how you can prepare it for when you get your bees. The exploded line drawing of a movable-frame hive in Chapter 4 of this book will help you better understand the way it functions. Put a paper clip on that page of the book for easy reference.

The major feature of a movable-frame hive is the ⅜-inch space on the bottom, the top, between the frames, and between the sides of the hive and the frames. This space prohibits the bees from gluing everything together with propolis (bee glue) as is their natural habit to keep their house tight against intruders. Those intruders are pretty well taken care of when you place the hive on a stand and close the entrance with cleats on cool and wet days. The bees *do* glue up any tiny cracks in the walls.

Almost as important as removing the frames easily is the straightness and uniformity of the frame and foundation. Left alone in nature, the honeybee will create glued-together, curved combs that are impossible for the beekeeper to extract the honey from. (See the picture of the wild comb Bill Robinson is holding in Chapter 2.) But with the manufactured frames and foundations, a beekeeper can remove each frame full of honey, place it in an extractor, and spin the liquid honey out.

*A 10-frame hive body with inner and outer covers and a bottom board.*

*A frame — one of 10 in a 10-frame hive body.*

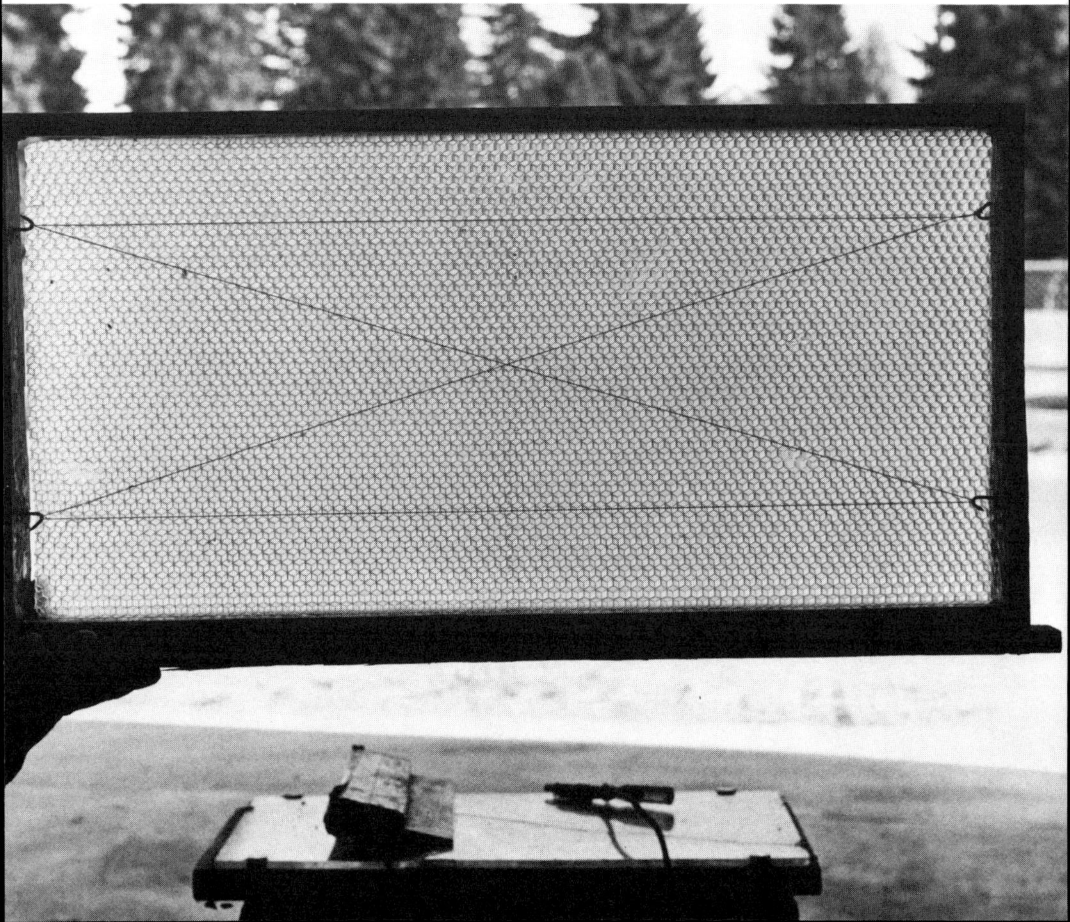

Another feature of the manufactured comb, or foundation, is that it saves hours of bee labor. Stamped in hexagonal cells, the foundation can be "added upon" by the bees. They do not have to start from scratch. They use the beeswax of the foundation — plus some of their own — to "draw out" or to deepen the cells, preparing them for brood and honey. Thus, man has devised yet another way to encourage the bee to multiply its store of honey. That extra honey, as we have said, belongs to the beekeeper.

The home you prepare for your bees will also need these essential items:

1. *At the beginning of the season*
   - One 10-frame hive body (or more, depending upon the number of hives you want to begin with)
   - One 10-frame bottom board for each hive body
   - One outer cover for each hive body
   - One inner cover for each hive body
   - Ten frames for each hive body
   - Wired brood-comb foundation for each frame (and several extra in case of breakage)

*A shallow super, a queen excluder, and a bee escape.*

2. *When you add the first and succeeding supers* (to be discussed later)

- Shallow super
- Ten shallow frames for super
- Wired comb foundation for each shallow frame (and several extra)
- Queen excluder
- Bee escape

These materials and those discussed in the following section are available from suppliers of bee equipment from various areas of the country. You may find one near you by looking in the yellow pages of telephone directories of large cities in your state or nearby states. Your library has these directories.

Ask for instructions for assembling the hives and frames when you purchase them. And after you have assembled them, place your hives (or hive) on the site you have chosen.

## Additional Equipment

***Protective equipment***

- Veil
- Helmet
- Sting-resistant gloves
- Smoker
- Bee suit
- Boardman entrance feeder (for no-nectar months)

***Extracting equipment***

- Hive tool (for removing frames)
- Uncapping knife
- Centrifugal extractor (if you feel rich; If not, use the less expensive ways of extracting your honey discussed later in this book)

Something else that will help you to learn how your bees think and act

*A variety of bee equipment: helmet and veil, smoker, sting-resistant gloves, hive, frames, bee escape, super, and entrance feeder.*

*Helmet, veil, bee escape, and uncapping knife.*

*An observation hive. Bees have formed comb on the inside of the glass.*

is an observation hive, which may be purchased from a supplier—unless you think you can build one yourself.

Now, after you have placed your hive or hives on your site, paint the brood box and the supers white with a tough outdoor paint. This gives the hive a more professional look, and it prohibits warping and rotting as the wood is exposed to the weather.

## When and How to Diseaseproof Your Bees

- If you have a county bee inspector, he will inspect your hive if you ask him to. Of course, if you have bought your bees from a supplier who has provided you with a certificate of the bees' health, this will not be necessary.

- To prevent American foulbrood, the most dreaded of bee diseases in the United States, add ¼ teaspoon of sodium sulfathiazole to each quart of syrup feeding 10 days apart during the spring and in the fall after harvesting the honey. Or you may mix together ¼ sulfa powder to ¾ powdered sugar and sprinkle the mixture over the tops of the frames.

The wax moth can cause as much damage to hives as disease, particularly in the southern states. The wax moth lays its eggs on the outside of the hives. The larvae are an inch long. They crawl into the hives and bore into the honeycombs, destroying them. A strong colony is the only known remedy for getting rid of wax moths.

Remember, as a hobbyist you owe commercial beekeepers the commitment to keep disease-free hives. Their businesses depend upon their bees, who can be wiped out with disease carried by a hobbyist's bees. Be sure you use the preventive measures described in this chapter as part of your beginning experience as a beekeeper.

You are now ready to start your first colony. Instructions follow in the next chapter.

*Wax-moth damage.*

# 8

# Putting Your Bees in Their Hive

Beekeepers vary in their opinions about whether to begin with one or more than one hive. I began with one hive because I had read one beekeeper's view that a new beekeeper is busy enough learning to take care of one hive the first year. Other beekeepers say it is safer to start with a few hives because if the queen of one hive dies, the beekeeper has at least one other colony left with a queen.

I guess I was lucky. My queen lived, and I walked on air when I first saw eggs and larvae in the cells of the foundation. That colony flourished through its first season and came through the winter beautifully. The second season was a lean one, though — through no fault of mine or the bees. After a good spring honey flow, the heat began. We had temperatures hovering around 100°F. from June through September with no letup and no rain. A second honey flow never appeared. Not even wild flowers did well that summer. I had to feed my bees sugar water, and I often had to sprinkle pollen into their hive to keep enough of them alive to come through that second winter. Commercial apiarists sang a sad song in Utah. But those who had seen the handwriting on the wall trucked their bees away early in June to cooler, moister bee yards in other states and saved their harvest of honey.

Whether you decide to have one hive or more, the process of installing your bees and taking care of them is the same. We're not going to talk about hiving a swarm or getting bees from someone in your vicinity.

*One hive to take care of may be enough for you in your first year. (Supers will be added later.)*

Hiving a swarm is too scary for most beginners, and getting bees from neighbors is too chancy — their bees might be diseased.

The best choice is to buy packaged bees from the South or from California. You know these bees have been inspected for disease. You may even request a certificate from the supplier to that effect. Moreover, packaged bees are easy to install. The only thing you have to decide is whether to buy Caucasian or Italian bees. You can't go wrong with either.

The Caucasian bee is gentler than the Italian bee. But the Italian bee has two advantages over the Caucasian: 1) it is hardier and more resistant to foulbrood, and 2) it does not use excessive amounts of propolis (bee glue) as the Caucasian does. Too much propolis between, around, and over the frames makes removal of the frames difficult. But remember, if you plan to keep your bees in your backyard, you may want the gentler Caucasian in deference to your neighbors.

When you have decided which kind of bee you want, the next thing you need to know is how to order them. Most bee supply houses will order packaged bees for you, and you either pick them up at the supply house or have them sent to you by express mail or parcel post insured. I ordered my Italian bees from Miller Honey Company in Salt Lake City and picked them up there on a prearranged date.

When the packaged bees arrive, they will be in a wire-netting cage, along with a can of sugar syrup and a mated queen in her own cage. Illustrations will also be provided, showing graphically how to do each step. Briefly, so that you will have an idea what happens, here are some instructions that will help you understand what you will be told to do.

Take the package to a cool, dark room until evening. But first, immediately paint the sides of the cage with sugar water (1½ cups sugar dissolved in 1 cup hot water) until the bees become calm. Their can of sugar water might be empty; mine was. Occasionally paint the cage with the sugar water until the time comes to place them in their brood box. They will be gentle and easy to handle if they are full.

In the cool evening, begin the process of hiving your bees.

1. Have an entrance feeder filled with sugar water ready to insert at the entrance of the hive. (Twenty pounds of sugar will be used by a 2- or 3-pound package of bees until they have gathered enough nectar for brood and honey.)
2. Have the brood box ready and sitting on a bottom board containing about five or six frames of foundation (leaving room for the package with a few bees left in it later). Have handy an inner cover and a top cover.
3. Just before placing the bees in the hive, paint the cage once more with sugar water.
4. Jostle the bees to the bottom of the cage. Some smoke from the smoker will make the bees more gentle. Pry out the feeder can and lift the queen's cage from the package.
5. Remove the cork from the queen's cage. (You will see an inch or so of sugar candy that the worker bees will have to eat through before the queen can be released. By the time the workers eat through the candy and reach the

*An empty queen's cage, with some of the sugar candy remaining at one end.*

queen, she will be familiar to them and they will not kill her.) Place the uncorked queen's cage in the center of the hive between two frames, near the top, with the candy end toward the bottom.

6. Now turn the package of bees upside down over the queen's cage in the hive and jounce and shake most of the bees out of the package. They are gentled and will let you do this. The queen requires many bees to be with her at this time so that she will not become cold. Place the package containing the remaining bees in the space provided by having left out a few frames. The package should be upright, with the hole at the top.

7. Immediately place the entrance feeder full of syrup

at the entrance on the bottom board and carefully place the inner cover on the hive, trying not to squash any bees. (Did you remember to add sulfathiazole to the feed?) Place the outer cover on the hive body.

8. Reduce the size of the entrance with an entrance reducer (cleat) so that cool air and robber bees will not interfere with the bees getting a good start in their new home. Entrance reducers come in varying sizes so that you may stop up the entrance to the size you desire.

9. Keep the feeder full of medicated syrup, and don't open the hive for about a week to see if the queen has been accepted and has laid some eggs. (Eggs are tiny, white, and elongated and stand on end at the bottom of the cell.) Some larvae should also be in the cells. If you can't find either, wait a couple of days and look again. If nothing has happened, the queen is probably dead; you will have to get another one right away. Success occurs more often than failure, however.

10. When pollen is not available to the bees in nature, shake some pollen over the frames about once a week until they can collect their own. Remember, pollen is vital protein for the brood.

11. When you have determined that everything is going well in the hive, remove the empty bee cage and replace it with the remaining foundations in their frames. The bees will go right to work on them, filling them with honey, bee bread, and brood.

You are now the owner of a colony of bees. Exciting and pleasant times are in store for you. One of those exciting times may be when you decide to move your bees to an area of more productive nectar flow. I hope it won't be so exciting as it was for my husband and I. I'll tell you about our experience in the next chapter.

# 9

# Moving Your Bees

If you are as lucky as I was in ordering and receiving a pamphlet on the nectar and pollen plants in your area, you will know whether it will be necessary to move your bees to a second — or third — honey flow.

In you decide to follow the honey flows, here is a way *not* to move your bees. I tell you this story because of the grief that came to my husband Bill and to me when we attempted to move our bees. Save yourself the same grief.

Novices that we were — and skittish about moving a brood box and three supers all at once — we planned to keep the bees all in one place and *away* from *us*: we would cover the entire hive with cheesecloth. That would let the bees breathe, you see, and it would keep them *contained*. My husband used to be a meat cutter, so he went to a meat packing plant and got a tube of the heavy kind of stretchy cheesecloth meat packers use for wrapping and hanging hams. That was our first mistake.

Our second mistake was to put that cheesecloth on the hive before the sun had set and the bees had settled down. We thought we were pretty smart. We got that tube of cheesecloth pulled down over the three supers and the brood box and under the bottom board that rested on the stand, and then we tied it off to the side. Of course, we were dressed in our protective clothing — gloves and veils included. Good thing: the bees were filling the cheesecloth, mightily riled — not only because we were disturbing them but because they hadn't been able to find any nectar for quite a while, and I had neglected to feed them before getting them ready to move (another mistake).

"How are you going to pick them up and put them on the truck without a dolly, Bill?" I asked, with perfect faith that he had the answer. After all, he had built a home all by himself, solving unsolvable problems all along the way, and this was no different.

Replying, "I'm afraid I'm going to mash some bees, that's how," Bill inhaled deeply, got a firm hold on the bottom board with both hands, and lifted the hive — brood box and supers — off the stand.

I must tell you that the top super had nothing in it except the foundation frames. The bees hadn't found anything to put into it. Another mistake: we hadn't removed that empty super. It was excess baggage.

Moreover, Bill hadn't been able to back the truck near the hive because the hive sat on a rocky slope that descended into a rockier gully that ascended again to a rocky piece of scrub-oak landing. His problem was to balance and carry that cumbersome 90-odd pounds of hive, honey, and angry Italian bees down the rocky slope and up again to the truck. Finding footing to carry *myself* wa all *I* could do as I followed Bill.

He almost made it. Near the top of the gully the hive began to lean to Bill's right. I didn't see it in time to help balance it. Before either of us knew what had happened, the hive had slipped from Bill's hands and wedged itself between a scrub oak and a rock. It was wedged *securely*. And the noise those bees were making was threatening. I got out of there.

But I turned in time to see Bill beginning to slice at the cheesecloth with his pocket knife. My frantic "No!" coincided with his bellow as bees boiled out the opening and attached themselves to him. By that time we were both running with bees fuming all around us. You see, bees come to movement (remember that), and we were really moving!

After three laps around the house and one lap through the scrub-oak gully on the opposite side of the house, we had lost most of our pursuers. Somehow I hadn't been stung, but Bill had — once through his glove (we had bought the least expensive kind, not the most bee-resistant kind) and several times across his back and arms — through clothing that should also have been a little more sting resistant.

"Why did you *do* that?" I exploded.

"I thought if I could just get the hive back to the stand in sections," Bill began defensively.

"You *never* mess with bees that mad," I groaned.

Later that night, after dark—though it was still very warm—we again dressed in our protective clothing. Except that Bill didn't bother to wear boots to tuck his pants into. We went out to see if the bees had calmed down enough for us to cut the rest of the cheesecloth off so that they could at least reenter their hive. We did get the cheesecloth off, and, feeling bolder, we decided to try to move the hive onto the truck. Bill set the supers aside, then instructed me to help hold the bottom board up against the bottom of the brood box as he lifted it. His last mistake for the day was to grasp the brood box from the *front*. Though it was dark, I knew what had happened as he let go the box and began to stamp around growling and shouting. Hundreds of guard bees had shot out of the entrance and clamped themselves to his ankles.

The next day Bill went off to work with a swollen hand, swollen feet, and a fever from countless bee stings (which have no doubt provided him with a greater resistance to their venom). So I called Joel Wright, who came with his wife that night and set the hive back onto its stand. After I had fed the bees for a couple of days and they had settled down, Joel and his wife came after dark, after it had cooled down, and moved the bees for us with an ease born of experience.

As I said in an earlier chapter, knowledge takes away fear. So now let me tell you how it *should* be done so that you needn't be afraid to move your bees.

1. Place you hive near an alley, or a driveway, or a flat piece of land where you can use a dolly to roll the hive to your truck.
2. Make sure your bees have been well fed for at least 2 days before you move them. This calms them down and gives them work to do so they'll ignore you.
3. Wait until the cool of the evening—until midnight if you have to. It's better to have to use a flashlight to see what you are doing than to disturb the bees while they are still active.
4. Never, never, *never* work from in *front* of the hive. If guard bees so much as feel a bump against their hive before they've settled down for the night, they'll dart out and attack whatever is outside their entrance.
5. If the bees haven't glued their boxes together with

*If it is necessary to move your bees in the daytime, wire netting across the top of the hive will allow for better ventilation*

propolis, tie ropes around them securely, or use hive staples so they won't slip.

6. Some beekeepers tack wire netting over the entrance to keep the bees in. But again, that's a dangerous place to work. And on the night Joel moved our bees, he didn't tie the boxes together, or staple them, or put wire netting over the entrance. The bees stayed in the hive for the 20 miles we moved them.

They didn't even show themselves at the entrance when Joel lifted them down and placed them on their new site. After learning how Joel did it, Bill and I were never afraid to move our bees.

### *How far should you move your hives?*

Since bees will return to their old location if they are moved only a short distance, they should be moved at least 2 or 3 miles. Foraging bees have become so familiar with certain guiding markers within that distance that they will always find their way back to where you originally had them. But if they are moved several miles, they will orient themselves immediately to their new surroundings and will begin foraging for nectar and pollen.

In their new location the bees will begin to memorize their surroundings and will go to work bringing in the welcome nectar.

So remember, when you must move your bees to provide them with the nectar and pollen they need, just keep cool, and keep your bees cool and well fed. They'll bless you for it with extra honey.

# 10

# Making More Room for Your Bees

Bees are not dormant in a hive during winter. They spend the early part of winter simply staying alive on the honey they have stored and clustering to keep warm. But toward the end of winter the queen begins to lay eggs in the cells, and the nurse bees feed the larvae with stored pollen and honey as the larvae develop into pupae. Because the nurse bees soon begin to need additional stores of pollen and honey, the field bees fly from the hive to forage on the first sunny day of spring.

In the early-spring buildup season, before the honey flow, the field bees gather nectar and pollen, the queen lays many eggs, the nurse bees feed the brood, and the hive seems suddenly to burst with bees. Sometimes, before a beekeeper knows it, the bees are swarming. When swarming occurs, the queen leaves the hive with at least half the bees. Only a nucleus (queen cells, brood, a few drones, and worker bees) remains. The honey harvest from that hive for the season will probably be only enough for wintering the bees. If the beekeeper is lucky, of course, he can capture the swarm and hive it.

## First-Year Supers

That's what happens to your bees if you've had them through the winter. But since this is your first year of beekeeping, the spring population of your hive will probably be more gradual. However, you should watch the brood box (hive body) carefully, especially if spring comes early and is accompanied by a few warm rains that bring on a heavy honey flow.

You don't want to add more room (supers) for your bees until they have nearly filled the frames in the brood box with capped honey. And you should also decide ahead of time whether you will use "deep" or "shallow" supers. A deep super is exactly the same size as the hive body or brood box. The good thing about this super is that the frames are the same size as those in the hive body and, therefore, they may be interchanged. But a drawback of a deep super is that it holds about 80 pounds of honey — making it a real challenge to lift. A shallow super is easier to handle, and you can exchange its frames with succeeding shallow supers. I chose "shallows" because I'm not very big.

After you have made up your mind which size to use, the trick is to know when and how to add each super.

***When to add a super***

- Do not add a super if frost is expected within a few days. Cold air space in the hive may keep the bees from maintaining brood-rearing temperature (90 to 100°F.), and many of the brood will die. The worker bees can be seen carrying out the white bodies of the dead brood. Also, the queen will stop laying eggs if it is too cold.
- *Do* add another super at the beginning of the honey flow. When foraging is good, bees bring in nectar and make honey so fast that a beekeeper is amazed, especially considering that the honey in the cells is half water that must be fanned away by the bees' wings before it is capped. The bees need at least two supers at this time. Check the brood box before adding a super to make sure at least 8 of the 10 frames are almost filled with honey and pollen. Otherwise the bees may begin filling the frames in the super before they have completed those in the hive body.
- Do *not* add a third (or fourth) super if the honey flow begins to wane. Otherwise the bees will only partially fill the supers. And full frames of honey are more desirable for extracting than half-filled ones.

*Add a second super at the beginning of a heavy honey flow.*

In your first year of beekeeping you may have to add only one deep super or two shallow ones, especially if you have started with packaged bees or a nucleus. It could take most of the season for the colony to build itself up to normal hive strength and to put away enough honey in the hive and deep super (or two shallows) for the bees' own use in winter.

*How to add the first super*

- On a warm day when you bees are bringing in nectar, dress in bee veil, bee suit (or other protective clothing), and bee gloves. Remember, there is nothing to fear if the bees are happy. And if your motions are calm and slow, they will pay less attention to you.

*Wear protective clothing when you add a super.*

- Raise the outer lid of the hive and give the bees a couple of puffs of smoke. Lower the lid for a few minutes and remove the outer and inner lids. Smoke the bees down into the frames, and place a queen excluder over the top of the brood box.
- Place a super with its 10 foundation frames over the queen excluder. If the super is deep, take four frames of honey from the hive body (brood chamber) and swap them for four foundation frames from the super. Place two honey frames at each side of the super. The bees and honey will lure the worker bees from the hive below. If the super is shallow, the bees will come up when they feel really crowded below.
- Place the inner and outer lids on top of the super, and let the bees go to work.
- If the weather is warm, be sure the hive entrance is not blocked with any cleats. The bees in both the brood box and the super need lots of room to come and go.

If the honey flow is strong, add a super every 20 days or so. The second super should be placed below the first super (between the brood box and the first super) so that the bees will store pollen in it, next to the brood chamber.

Be sure to leave either two shallows or one deep super besides the brood box for yours bees to use in the winter. Any additional honey is yours.

# 11

# Harvesting Your Honey

The title of this chapter *might* prove to be a joke or at least a postponement of satisfaction. It depends on how well you have followed the instructions in this book and how well nature has supplied your bees with nectar and pollen.

So if you knew something about beekeeping before you started, if you found a suitable site for your bees, if you prepared the hive and successfully installed your bees, if you moved them to the best honey flows you could find, if you put the supers on at the right time, and if nature cooperated — after all these things have been accomplished, you *may* harvest *your* honey. Not the bees' honey — *yours*.

And when a new beekeeper extracts his first honey, a special glow surrounds him. It's a glow of pride in his achievement. Yes, it's true the *bees* made the honey. But they wouldn't have made so *much* without the beekeeper's help. So pat yourself on the back.

But you have some more to learn before you can sit back and feast your eyes — as well as your stomach — on all your jars of golden honey.

## When You Should Take Off Your Supers

As I have said, when your bees make more than one deep super or two shallow supers of honey, the remaining honey is all yours. Sometimes in the first year you find only a few frames of honey in the extra super. Just smoke the bees off these frames, take the frames into your kitchen, and simply scrape off the honey — wax and all — from both sides of the foundation, using a spoon and being careful not to damage the foundation, so you can use it again. Put the honey into a metal or plastic bucket or jar and it's ready to eat. It's great on hot biscuits or on hot homemade bread. The heat melts the wax, and the honey is delicious. It's not necessary to buy an extractor for such a small amount of honey.

Take the foundation, sticky with honey, into a cool dark place, wrapped in plastic to protect it from mice and pests. Store it there until next year; then use it to put into a super to get your bees excited about starting a bang-up year. If, however, you have removed this frame while the honey flow is still on, put it back into the super and let the bees get back to work.

When you harvest honey and you find you have one or two extra supers full of honey for yourself, be sure you take the supers off at the right time:

1. Remember, when bees are busy working, they will pay little attention to you. Be sure the wind isn't blowing and that the sun is shining. Fewer bees will be in the hive, and it will be easier for you to take out the combs of honey.
2. Take the supers off during the warmest part of the day so that the bees won't have to work hard to heat up the hive again. If you interrupt their work, they will return less honey.
3. Two-thirds or more of the foundation in the super should be capped. This will show that the honey is ready and sealed up.
4. You may wait until the honey flow is finished before taking off your supers. But if you do this, be sure to continue to add enough supers during the honey flow to keep the bees from swarming.

# How To Extract the Honey

Your first step toward extracting liquid honey from the combs is getting the combs out of the super (or supers).

### *Taking the combs from the super*

- Wear a veil, a bee suit, and gloves. Smoke the bees to gentle them.
- If there is more than one extra super, take all of them off the hive and place each one on a bottom board. Place a cover over each one. Replace the super that is *least* full of honey.
- Place a bee escape over the hole of an inner lid and set this inner lid over the super least full of honey.

*Butyric acid ("Bee-Go"), as the label on the bottle says, "chases bees from supers."*

- Place the full super or supers on top of the bee escape. It may take a day or two for all the bees to leave these full supers through the bee escape. Then you can remove the supers and their foundations full of honey. Some beekeepers smoke the bees down from the supers, but this might oversmoke them, and they will not be able to work very well. Other beekeepers brush the bees down with a bee brush, but sometimes this angers the bees until they sting. Many commercial apiarists use butyric acid on a pad over the frames. But if you use butyric acid, the odor will be so offensive you can hardly get near it, even though the bees have rushed from it to lower supers.

  Bee escapes take longer to work, but they are probably the most desirable way to remove bees from the supers — if, of course, you have the time to wait.
- After you have removed the supers and foundations, remove the bee escape and, later, if necessary, set a replacement super over the super remaining.

### *Taking the liquid honey from the combs*

- Have everything ready ahead of time.
  1. You will need an extracting room that keeps bees out.
  2. You will need a tub or sink to catch wax cappings. A large, round basin with a bar resting across the top to steady the frame of honeycomb would be convenient.
  3. If you want to invest the money, a two-frame, hand-operated extractor can be used. (Or find a beekeeper in your vicinity who will extract your honey for a small fee.)
  4. This is important: don't try to fill jars directly from the extractor. They will run over and cause a big mess. Instead, fill clean 4-gallon cans, from which you later can pour the honey.
  5. Heat an uncapping knife. If it is nonelectric, wipe the hot water off before using it; water helps ferment honey.

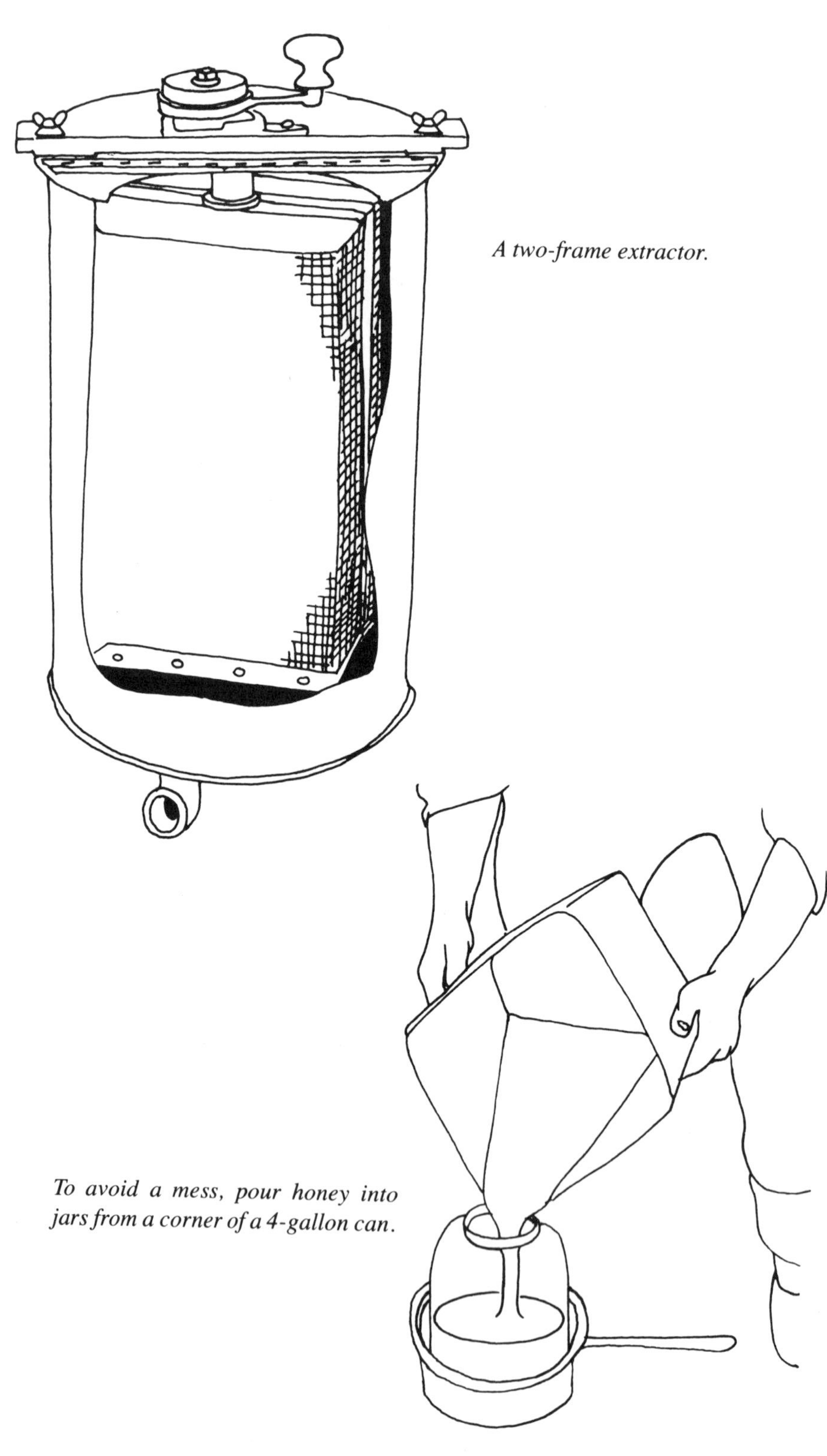

*A two-frame extractor.*

*To avoid a mess, pour honey into jars from a corner of a 4-gallon can.*

- Extract the honey
  1. Uncap the combs by resting the frame on the bar across the basin, leaning the frame forward so that the cappings will fall into the basin. Work the hot knife back and forth from the bottom of the comb to the top. Do this on both sides of the comb.
  2. Place combs of equal weight opposite each other in the extractor. Turn the crank slowly a few times. Reverse the combs and do this again, then once more, until all the honey has been extracted.
- The combs will have some remaining honey on them, but place them directly into a super. If the honey flow is still on, place the super on the hive box to attract the bees from below. If the honey flow is past, store the super in a cool place so it is ready to be used in the spring.
- You can spin the honey out of the wax cappings by putting them into capping baskets in the extractor and turning the handle fast; or you can drain them from a cheesecloth wrapping into the 4-gallon can. After you have washed and melted the cappings, you may use them for capping jars of jelly or jam.
- Pour the honey from the corner of the 4-gallon can into jars, using a funnel if necessary, and place lids on the jars. Store them in a cool place.

Clean your extracting room. Then enjoy your honey. You have earned it.

# 12

# Tucking Your Bees Away for Winter

If you live in a place where the winter temperatures never rise above freezing in the shade, you may experience trouble wintering your bees. But if you live where there is a thaw or two during the winter, you have a good chance to keep your bees alive until spring. Some beekeepers do not wrap their hives if they have a thaw or two because wrapping keeps them warmer and thus more active, and they eat their winter stores too fast. However, if you are nervous about your bees freezing, the following suggestions should help.

## Leave Your Bees Plenty of Honey for Winter

For winter, a strong colony of bees should have a super or supers containing at least 70 pounds of honey — besides the hive body, which will contain 20 to 30 pounds of honey. And it is very important to leave four to six frames of pollen in the brood chamber. Remember, the queen begins to lay eggs before bees begin their field work, and the brood will need the pollen as well as the honey.

In cold climates, at least six combs in the brood chamber should be smothered in bees, ready to be wintered, by the middle of October.

*When at least six combs in the brood chamber are smothered in bees, the bees are ready to be wintered.*

# You Might Pack Your Hive

If you decide your winters are colder than one- or two- thaw winters, you should consider one of the following methods to keep your bees warm enough to spring.

1. Your bee supplier can sell you *winter cases* that fit over your hives. These are quite expensive, so you may prefer another method.
2. Black tar paper

- Black absorbs heat and warms the colony on a winter day, allowing the bees to take a flight to expel their feces.
- The black tar paper helps melt the ice and snow from the hive's entrance so that the hive is properly ventilated.
- The tar paper protects the hives from the elements, reducing deterioration.

When you pack your hive with tar paper, follow these instructions:

a. Cut a piece of wood ½ × 2½ × 1½ inches. Drill a small hole (⅝-inch diameter) in it and set it aside. This will be the top entrance to the hive.

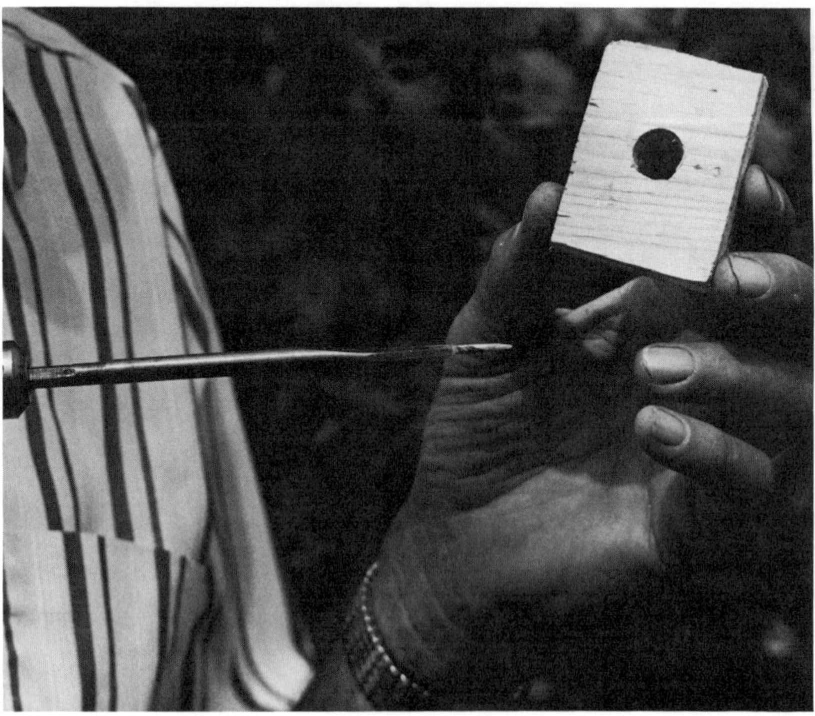

*A small entrance drilled in a piece of wood serves as the bees' top entrance in the winter.*

*Tacking tar paper onto the hive.*

*Nailing the top-entrance block of wood above the handhold in the super.*

b. Seal off the lower entrance to the hive. It will become blocked with snow and ice in the winter. A top entrance is preferable.

c. Drill a ⅝-inch-diameter hole in the small top-entrance block of wood just above the handhold in the super.

d. Wrap the tar paper around the hive (both the lower and the upper story), marking the place to cut the entrance hole through it. Tack or nail the tar paper onto the hive. With a sharp knife cut the entrance hole through the tar paper.

e. Nail the top-entrance block of wood over the hole in the tar paper and the hive. Your bees' winter entrance is now ready.

f. Place an 8-to 12-inch stack of leaves over the inner cover. Place the top cover over the leaves, cover with tar paper, and tack it down.

g. Bees give off moisture, which you must help them to get rid of. So drill a ⅜-inch hole above the sealed lower entrance for dry air to enter and carry moisture out the top entrance.

3. Securing *dried leaves* to the sides and back of the hive with wire netting can also keep a hive warm, though not as well as black tar paper. With this method the front of the hive is left open, and the lower entrance (made small enough with cleats to keep out mice) is used. Again you will have to keep the entrance clear of snow and ice.

Once your bees are ready for winter, do not open the hive. Inspect it once in a while for wind damage (or animal damage) and make any necessary repairs. Check the entrance hole to see that it is not clogged with ice or dead bees. The cold weather always kills some bees, so don't be alarmed if you find some dead ones around the hive. Bees cluster tightly in the winter to keep warm, and those in the center of the cluster (except the queen) take turns moving to the outside of the cluster and pushing the cold outside bees to the center for food and warmth. In spite of this democratic process, a few are always lost.

Winter is a good time to get ready for spring. Build or buy new hives. Get your frames ready. Get ready to feed your bees sugar syrup if the winter is long. Your second year of beekeeping will soon arrive.

PART III

# Beekeeping For Profit

# 13

# The Game of Commercial Beekeeping

The biggest honey producer in the world began his business at the worst of times. Woodrow Miller started keeping bees commercially at the height of the depression in 1934. He had worked his way through college and wanted to go into honey producing with his brothers and his father. The Department of Agriculture in Washington, D.C., told him beekeeping was the worst business he could enter, with one exception: raising goats. The methods followed by the Millers in achieving success in commercial honey production despite that warning are revealed in this and the next three chapters.

If you are interested in learning how to be a success in the business, Miller offers some suggestions:

1. *It is most important to work at least one season with a commercial beekeeper in the vicinity where you would like to set up business.*

"It's a different world," says Miller, "keeping bees in Hawaii than it is in Utah or in the desert or in Alaska." You have to learn many things, such as whether the area contains little water or an abundance. Bees require a great deal of water to produce honey. You should also know whether or not you need to wrap your hives in the winter. If you wrap them where the weather is mild, the bees remain warm and active in their hives and eat their winter's store of honey too fast. If you work with a

commercial beekeeper, he can advise you about the most profitable method of wintering your bees.

A commercial beekeeper can also tell you about the nectar and pollen plants in your area and where to move your bees for maximum honey flow. He can tell you what poisonous sprays are used and what you should do about them. He can answer all your questions before you need the answers.

And you may learn, in that season, whether or not you want to continue in the bee business.

2. *Find good locations for bee yards.*

Southern California may be a paradise, but as an area for producing honey it is a "no-good," according to Miller. Although he lives in southern California, Miller does not have one colony of bees there. It is a desert, lacking water and sufficient nectar. His most productive bee yards, operated primarily by his nephews, are located in the midwest: Nebraska, Montana, South Dakota, Minnesota, and parts of Utah — though water is limited in Utah.

One honey producer in Provo, Utah — Joel Wright — gives some good advice about finding productive bee yards. He says that the yards are never the same from year to year. A farmer might plant alfalfa one year and grain the next. Find a place for your bee yard that will be within a 2-mile radius of more than one crop. Then the bees will not depend on only one crop.

The mountains of Montana are in a high-moisture area. Sometimes an apiarist can drive his truckload of bees to a place where the clover is so high it hides the truck. Here every hive produces 100 pounds of honey in 3 weeks — compared with an average of 35 pounds per hive in Utah.

You must also learn where to move your bees for the early spring buildup and for pollen. This is the time when the brood is being fed "bee bread" — a combination of honey and pollen. Without pollen, the new little bees could not develop into efficient workers.

Joel Wright takes his bees to northern California in the early spring "strictly for the spring buildup and for pollen." Then he returns them to his yards in Utah.

Knowing where the best bee yards are can make all the difference to a successful honey-producing operation.

3. *Plan to operate large enough units of bees for successful honey production.*

"It used to be that a quarter-section farm was a good farm and would



← *A field bee gathering nectar.*

*Nectar plants: clover, a favorite of bees.*

→

*An empty queen cell and capped brood, all made possible by sufficient pollen for the nurse bees to use.*

raise a family,'' says Woodrow Miller. ''No more. It takes a section of land now.'' With honey production, according to Miller you had to have 500 colonies of bees to make a living. ''Now you should plan on a unit of from one to two thousand colonies.''

Remember, 1. get some practical experience with someone who knows the score, 2. find a good location, and 3. operate large enough units, and you can make a good living in the honey business.

# 14

# How to Finance Your Business

"I have roughly $150,000 to $175,000 worth of bees and can't borrow a cent on them," says Joel Wright. That's not surprising, considering a "banker cannot put the bees in his vault," as Woodrow Miller says. So how do you finance a business in beekeeping if you can't use the bees as collateral (insurance against nonpayment) on a loan?

Remember, you will need 1,500 to 2,000 colonies of bees for a successful business. That means that for each colony you will probably buy a brood box and two deep supers with bees and honey. The price for one such colony ranges from $45 to $100, depending upon the strength of the colony (and the whim of the apiarist). If you paid somewhere between that—perhaps $75—your cost for 1,500 colonies would be $112,500; for 2,000 colonies it would be $150,000. That's a lot of money.

You will also need equipment to operate your business successfully—pallets, uncapping machines, lift trucks, extractors, feed tanks, storage tanks, and other items discussed in the next chapter. This equipment will cost thousands of dollars more.

Here are some ideas for you to consider if you are serious about financing a beekeeping business.

*Some equipment used in commercial beekeeping: a 300-gallon feeder tank, a fork lift, a truck, and a trailer.*

*Sixty-pound plastic buckets of honey and six-pound cans.*

# Begin as a Hobbyist and Grow Gradually

If you start now, you will have years to build up your bee business. The know-how to obtain maximum honey yields each year will come with experience. You will also be better able to deal financially with frosts and droughts because you are paying for your business as you go, and you won't have a loan payment draining you each month.

As a hobbyist you will learn how to keep your bees free from disease, how to increase your existing colonies, how to keep your bees from swarming, and how to move them to productive nectar fields.

Also, if you can curb your generous nature and not give to friends *all* the excess honey your bees produce, the money you make each year from selling honey can be spent for more colonies.

*Joel Wright shows two young people how to place a frame of honey into an eight-frame reversible extractor.*

During these building years, you may have to make a deal with a commercial beekeeper, for a minimum fee, to move your colonies to nectar fields with his equipment. You may even earn the money for that fee by renting your colonies to a farmer or a fruit grower for pollinating their crops. Remember, a good honey crop depends on moving your bees to where the nectar is. It is not possible to harvest a big crop if you keep your bees in one place.

Or, if you are lucky, your dad or an uncle might rent a lift and a truck and help you move your bees. If you have a real desire to build your business, you will find the way to do it. Find out what a commercial beekeeper charges to rent his equipment. Then go to work on your relatives or friends. They'll help you move your bees for a price — say a can of honey. It will be well worth it.

As for harvesting your honey each fall, you will have to decide whether you will invest some of your profit in an extractor or whether you will rent the use of a commercial beekeeper's extractor. Ask yourself which will be the most profitable for you — buying more colonies or an extractor? How much profit will the new colonies yield — enough to pay the rent on the extractor, with some to spare? Probably not immediately. But within a year or two they might. The decision will be yours. And making decisions about your own business is, after all, the stimulating part of being an entrepreneur.

## Lease Bees from a Commercial Honey Producer

According to Woodrow Miller, most beekeepers in the depression years "did not make it and went back to teaching school, being a carpenter, or whatever they had done before to make a living." Obviously, the Miller family was not among those beekeepers who failed. I asked him how he accounted for their success, considering that the mortality of beekeepers was so great. He replied:

> My father worked 24 hours a day and had very basic policies: five of us boys all worked, and he divided his business at that time among his sons and his employees. He leased them units of about 2,000 or 3,000 colonies, and they operated as though they owned them. And

when they could buy half the unit, he would sell them that half and lease them the other half. They would continue to operate them just like their own. They could get Mama and Papa and relatives and neighbors to help them harvest; my father didn't have to do that work. And he didn't have the wages and expenses. In that way he was able to get successfully through the depression, and the others were able to make a living. It worked out fine. We still use that philosophy.

Does the Miller philosophy give you some ideas?

- Remember, one way to become a successful beekeeper is to work a season with an established honey producer. If you work hard and prove yourself a capable beekeeper, why not ask your employer to lease you some bees? You will take work off his hands, and he will receive a percentage of your profits. You may want to start small and in a few years work up to the maximum number of colonies for a good living.
- Get your relatives and friends involved. At first, you will probably be going to school and will not be able to work full time among the bees. Often, simply for fun, your close friends and family members will assist you in feeding and moving the bees and in harvesting the honey. A reward of a small can of honey may suffice for their efforts. Their greatest reward, however, will be the education and the excitement they derive from helping you.
- When, over a period of years, you have built up your number of colonies into a large operation through leasing and buying, remember how you got started. Give others a chance, as the Millers do, to get started in the business. It will benefit you financially; but better yet, it will provide a way for you to make a contribution to others.

# Borrow Money From a Lending Institution

Are you a gambler? Do you enjoy the thrill of taking a chance? Then borrow money to start beekeeping on a small scale. As an example, say you are older and married. You already have a pickup truck, and you want to start out with about 100 colonies while you work a full-time job. You want to have a fork lift, a dolly, an extractor, some pallets, and a large feeder tank. You estimate you can obtain all these for about $10,000. All you have to do is have that much money invested (an equity) in your home and your personal property. And of course you must be able to find a lending institution that's willing to make the loan — not always possible.

*A feeder tank with pump being loaded by a fork-lift onto the bed of a truck.*

Tell the lending people you've learned the technical end of keeping bees (you can get a reference from your employer to that effect), and you'll mortgage everything you have. The magic word here is "mortgage." Lending institutions want collateral—especially equity in real estate (your home and land) and personal property (for example, your pickup, your car, your boat).

Keep remembering the reason you can't borrow money using your bees as collateral: they can become diseased and die in a matter of weeks; they can die from the cold in winter; they can die because of drought; and they can die because of a late frost that cuts off the honey flow.

Each of these problems can usually be handled if one occurs without all the others—and if you are financially stable. The magic of antibiotics as prevention against disease works well, if done properly. Packing the hives against an unusually cold winter can save many bees (although it is an expensive and a somewhat time-consuming task). Providing a substitute water supply during a drought is often possible (but also somewhat expensive, if you rig up a circulating water supply). Feeding the bees sugar water when frost has taken the nectar can save many of them (but, again, this costs money).

The point is, if you do not owe money for the bees, the expense of protecting them against drought, freezes, and disease will not seem so great, If you do owe money for them and you lose a large percentage of them, you will be left "paying for a dead horse."

Remember: if you have mortgaged everything you own you can lose it all. But if the gambling fever is in your blood, if you love the excitement of taking a chance, if you don't want to wait until you have gradually and safely built up several large bee yards—either by expanding on your own or by leasing from an employer—then borrow money.

Again, the decision is yours. These three alternatives are for you to think about and to act upon. If you really want to become a commercial beekeeper, you can.

# 15

# Learning to Manage Your Operation

You've worked for a time with an established apiarist, and now you're pretty confident you can take care of bees. You've decided how you're going to finance your own honey-producing business, and you realize that you will have to make your bees produce more honey than they would in nature — more than to keep themselves alive through the winter.

Now your challenge is to make your business pay off. And that means you'll have to manage it properly. Somehow you'll have to make everything work together for your good: handling employees and their salaries, mechanizing, feeding and migrating your bees, keeping them disease free, diversifying — all the things that make an operation successful.

Again, you can let the experts help you. They've been through it. Why not profit by their experience?

## Take Some Classes in Related Fields

Joel Wright says a successful commercial beekeeper is a mechanic, an economist, an agriculturist, and a salesman. And he adds one more thing: he is a hard worker. Being a mechanic and a hard worker — even a

*Some courses in marketing, managing, and agriculture will help prepare you to become a successful commercial beekeeper.*

salesman — comes naturally to some people but most of them have to learn something about economics and agriculture.

When you're in high school, take a class in economics. Planning to become a beekeeper can make that economics course more meaningful — and therefore more enjoyable. If you attend college or a technical school, sign up for courses in marketing, managing, and agriculture. As you study these courses, think of the ways they can help you when you become a beekeeper.

Woodrow Miller says beekeeping is a *part* of agriculture. ''We have our own farms in Nebraska, and we have put some funds into agriculture to produce crops germane (necessary) to ours.'' His brother put money into crops and ''made a million dollars along with beekeeping.'' So you see, while you are producing nectar for your bees, you are also producing agricultural crops for your bees to pollinate. Then you can sell those crops on the open market for a profit. And you have assured your bees of a season of nectar.

## Learn How to Handle Employees

When it's time to get down to the nitty gritty of operating a business, you'll have to think about the labor needed to make your beekeeping operation a success. There is always work to do: trucking bees to another location, feeding, getting ready for spring, among other things.

### *Screening prospective helpers*

It isn't easy to find someone suited for working among bees. Here are a few things to consider before you hire help:

- Is your prospective helper interested in becoming a beekeeper?
- Is he afraid of bees?
- Is he a hard worker?

Joel Wright tells of an experience he had with an eager beginning employee who wanted to work with bees. ''Are you sure?'' Joel asked him. ''The bees are mean this time of year. We'll be doing some feeding and some riding through the bee yards, and they'll get unhappy with us for disturbing them.'' The applicant's reply was, ''No problem.''

The first day in the yard the bees became fairly mean. ''I have seen them much meaner than that,'' Joel said, ''when even I have been driven out of the yard.'' But his helper wanted nothing to do with it. He got stung once on the wrist, and that was it. ''He would have no more — wouldn't get near the hives.''

Joel says he gets used to being stung, and he works better without

*A commercial beekeeping operation requires employees who are not afraid to work hard. Using a hand truck will, of course, make the work go faster.*

gloves than with them. (Just make sure you're not allergic to bees.) Feeding bees in a no-nectar season — spring or fall — is no work for the faint-hearted. Test an applicant before you hire him. Find out if his fear of being stung is stronger than his desire to learn the beekeeping business.

### *Settling on salary*

- Find out if your helper is a hard worker. Will he want to work for minimum wages and do hard, hot, sticky work — even to learn the business for himself?
- Do you really need that helper? Joel Wright says that he can, with a lot of work, handle a maximum of 1,000

> colonies by himself. But he adds quickly that's because he is almost completely mechanized. Even with mechanization, you will need help to take care of more than 1,000 colonies.

Remember, you have to screen an applicant carefully for his desire to learn, his courage among the bees, and his ability to work hard, long hours.

## Mechanize

"With a large operation you need mechanical equipment. You can't do it all by hand," says Woodrow Miller. And Joel Wright agrees. The following are some of his mechanized equipment.

### *A 300-gallon feed tank*

"Most beekeepers," Joel says, "do not invest that kind of money for a big feed tank. But where I can feed 500 colonies in a day, they can feed only 50 to 75." Joel's feed tank has a honey pump connected to it — 50 feet of hose with a gas nozzle attached. The tank is on a truck, and "we just drive around the apiary feeding the bees." Joel uses division board feeders — the kind that sit down in the supers and hold up to 2 gallons of sugar syrup.

He told me an interesting reason he doesn't use entrance feeders (other than that they hold less than the division board feeders): "The entrance feeders (small jars) are prime targets for B-B guns."

Bees should be fed in the fall rather than in the spring because it stays warmer longer in the fall. Often, in the early spring, it is too cold to feed them.

### *Loader*

This mechanical gadget has a fork lift to transport the heavy hives to the bed of a truck. Once loaded, the hives can be moved to another location flowing with nectar.

*Joel Wright uses a 300-gallon feeder tank with a 50-foot hose and a gasoline-type nozzle to feed his bees. Here he demonstrates with an empty hive.*

*An electric uncapping machine. The frames are uncapped in this manner before they are placed in the extractor.*

### *Pallets*

When the hives sit on pallets, the fork lift can scoop them up more easily and lift them to the bed of the truck. "If you don't have a loader, you're going to have a bad back. You can plan on that," says Joel. "Sooner or later you're going to lift a hive heavier than you should lift. Back fusions and hernia operations are common among beekeepers." Beekeeping is hard, heavy, long work. And a loader and pallets can cut down on that work. Plan for them as some of the necessary pieces of equipment in a large operation.

### *Extraction equipment*

A large extractor will spin out many frames of honey at one time. Look at many commerical extractors, comparing their capabilities and prices, and get the one best for you.

### *Trucks*

A good-sized yard of hives full of honey requires plenty of transportation equipment — not only for feeding and moving the bees but also for getting the finished product to the market. Ask commercial apiarists what kinds of trucks are the best for your needs — and how many are absolutely necessary. An apiarist near you may be eager to pool his trucks with yours for the business of moving bees and transporting honey to market. Share and save expenses wherever you can.

## Migrate

"You've got to be a migratory beekeeper to be successful. You cannot stay in one spot any more. That's a gone era. You have to pick up the bees, put them on a truck, and haul them," says Woodrow Miller.

And he should know. He has operations in Nebraska, Minnesota, South Dakota, the Colorado River, Utah, and the Palos Verde Valley at Blyth, California. Delta, Utah, is better for bees than northern Utah; but all of Utah has only about 80,000 colonies. On the other hand, Nebraska, South Dakota, and Minnesota each have about 200,000 colonies.

There are some problems you may have when moving your bees.

*A large can of honey can be an inducement to a farmer or a landowner to let you place your bees on his property.*

First, most people don't want bees on their property. They're afraid they'll sting their animals — and them, too. So you may have to pay for some of your bee yards. Promise the owners a 5-gallon can of honey. Sometimes they'll consider it. Joel Wright says he pays 45 cans of rent a year.

On the other hand, if you find an orchard of sweet cherries, each colony can bring you from $4 to $6 in pollinating fees. In an almond orchard in California, each colony can earn anywhere from $8 to $15. Orchardists, however, require colonies that are at full strength. And it takes a lot of feeding and preparing to get them to such strength. Eight full frames of bees and brood per colony are required in California by the first of February, "and we don't usually have that kind of strength so early in the season," says Joel.

So that's another challenging part of becoming a big-time beekeeper — keeping your eyes and ears open for places to move your bees where nectar flows plentifully.

# Sprays and Diseases

### *Sprays*

I asked both Woodrow Miller and Joel Wright what they did about orchardists and farmers who spray their crops. Miller replied that "the only thing you can do is cooperate with the farmer and grower and get them to use the proper timing and the less toxic materials — or move your bees away." He says he does a combination of both. "Most honeybees are smart," he says. "They stay away from most toxic sprays." But he adds there are some sweet and subtle sprays, such as arsenic (supposedly outlawed), that the bees take home; then the baby bees die.

Orchardists are quite cooperative because they realize how vital the honeybee is to the agricultural economy of the nation. "The alfalfa growers and the almond nut growers are cooperative. Almond growers couldn't produce *one* almond nut without the honeybee." So they cooperate with the bee industry by spraying at night when the bees are in their hives. Moreover, the bees don't get involved if the blossoms haven't yet appeared. "We work it out cooperatively," says Miller.

Joel Wright adds that not *all* growers are so well informed and cooperative. "You'd be surprised at how uninformed some of them are. But then, sometimes they get between a rock and a hard place. They think, 'If I don't spray, I won't have crops. So I take my chance. If I spray this bee man out, there will be somebody else to replace him next year.' "

When that happens, Joel Wright, like Woodrow Miller, loads his bees on his trucks and hauls them away from there. And it is the grower who suffers from lack of pollination for his crops.

### *Diseases*

For commercial apiarists, disease is a crucial problem. They must use every means of disease prevention possible. State laws are strict. A diseased colony must be destroyed; the frames must be burned, the bees killed, the boxes scorched. In the spring, the apiarist must give 8 to 10 doses of sulfathiozole in the bees' syrup before moving them to another nectar field. Whatever you do—don't neglect disease prevention. And if you know any hobbyists who are not treating their bees for disease, get after them. Their bees could infect yours.

## Diversification

Sometimes a business lends itself to diversification (sidelines that bring in money).

### *Beeswax*

This is really the only by-product of the honey industry. Beeswax may be used for many things: as a base for good cosmetics (lip rouge, vanishing creams); as candles for sacred sacraments in the Catholic Church—candles that must be 51 percent or more beeswax; in combination with paraffin, to dip fruits such as apples to preserve their luster; and of course, as foundation for bees.

When it has been bleached and refined, it is considered USP virgin (pure). Unbleached beeswax is about $1.75 per pound, and purified beeswax is about $3 a pound. An apiarist can salvage about 1 pound of beeswax from 100 pounds of honey.

Woodrow Miller says the interest he pays on his foundation plant and his beeswax refining plant "sort of takes the profit out of that aspect of it." That's a good thing to know.

### *Pollination*

As you have read, an apiarist can earn from $4 to $15 per colony from growers if he can build his colonies to peak strength by very early spring. This might be worth an all-out effort (feeding the bees into late fall, wrapping the hives in severely cold weather, making sure the bees have a plentiful store of honey for the winter—anything to build up the colonies before moving them to that important orchard or farm).

### *Agriculture*

We have discussed this subject a bit, too, but I feel it worth repeating here. Money is to be had in food crops. If those food crops are the kind

*To make sure your bees have plenty of stores for the winter, give them sugar syrup into the fall. (See large storage can at left.) Joel shows some of the syrup in a drum covered with a lid.*

that produce rich nectar for bees, the two complement (help to complete) each other so that a profit can be made from both.

### *Comb honey*

You may wonder why comb honey is mentioned here as something separate from regular extracted honey. The answer is *money*. "Producing comb honey is becoming a lost art," Woodrow Miller tells us. Beekeepers have to do a lot of detailed work to produce it, and they have to know what they're doing. But the price for comb honey is extremely high: "$20 a case for a dozen little squares of honey. The money to be made there is literally a gold mine. People demand it. Their doctor tells them to get pure, natural honey; it will help them with their sinus, their allergies; and of course you don't ask the price when the doctor gives you a prescription." Miller concludes, "The potential is great today."

The experts have shared with you some of their knowledge about learning to become a beekeeper, financing the business, and managing the operation. In the remaining chapter they will advise you briefly about marketing your product.

# 16

# How to Market Your Product

From being the "worst business in the country" (except raising goats) to one that is economically very encouraging, honey has established a good market for itself.

"In 1935, we were begging for a market for our honey at about 4⅝ a pound," says Woodrow Miller. "It cost a nickel a pound to produce it. Beeswax was about a dime. Today honey is from 37⅝ to 47⅝ a pound (on the wholesale market), and beeswax (unpurified) is about $1.75 a pound. So it's a completely changed business."

A valuable tip Miller gives is that much more money is to be made in honey production than in packaging and distributing it. In other words, get the people coming to *you* — both wholesalers and retailers — for your goods. And it's worth repeating that if you have comb honey, the market *will* come to you — in droves.

Joel Wright has marketed his honey in a "number of ways." When his father, a teacher, had four summer months vacation, he could harvest and sell his own honey for a good profit. But teachers no longer have four free summer months; and Joel's father is easing out of the honey business. It's up to Joel to market his own honey. "If I have a good year, I wholesale the honey to Miller's. I try to guess on the retail market."

Joel says he has never really pushed a retail market, but he has sold a lot of honey from his warehouse, just by word-of-mouth or by an ad in the paper.

*Honey is now "a gold mine," says Woodrow Miller.*

*A 60-pound plastic pail of honey.*

I know how that works. My husband and I used to go searching for a beekeeper with some honey. We'd ask friends who had bought honey; we'd look in the yellow pages of the telephone directory; we'd go to health-food stores. And eventually we'd track down a beekeeper who could sell us some "pure" honey.

Many people hunger for good, unadulterated honey. The market is there. All a beekeeper has to do is cater to that market. That may sound easy, but there are some pros and cons to getting that honey to the consumer. I'll list the cons first.

### *Cons*

- First of all, *labor* is a serious problem. Both Wright and Miller agree that most people don't want to do such hot, hard work for minimum wages. It's hard to find good workers.
- A good location for bees is usually a "lousy place to live." Miller says they produce honey on almost all the Indian reservations in the Middle West and western America because they lend themselves to honey production. "But nice, pleasant little villages in southern California, for instance, do not lend themselves to honey production." So it is difficult to find an area for producing honey that is also a good residential area.

Joel Wright, for example, loves living in north-central Utah. But, he says, in certain places of Montana a beekeeper yields twice as much honey as Joel does because Montana has more water and vast fields of nectar.

- It is a hazardous business. In Utah where Joel has his business, he's not certain of any income at all. A June frost can destroy a honey flow that will take 2 or 3 years to make up. A hot, dry summer can burn up a second honey flow.

Disease and wax moth are other hazards of the beekeeping business.

Why, then, do many beekeepers stay in business? What keeps them, year after year, bringing their golden liquid to the market?

### *Pros*

If done successfully, "beekeeping is a more profitable game now than it ever was in all of recorded history — literally a gold mine," says Woodrow Miller. "I have made a fortune out of honey production."

The potential is great today. And it is an exciting challenge to try to take advantage of that potential.

- "Beekeeping gets in the blood," says Joel Wright. "Most commercial beekeepers sing a dismal song, but I enjoy the challenge. I like working for myself; I like the job. There is real satisfacton in being a producer and a salesman. When you get that end product and put the

*If you get a thrill out of watching bees making honey, perhaps beekeeping is a vocation you would enjoy.*

> honey on the table, there is a satisfaction to be found nowhere else.''

Business of all kinds abound and attract applicants of all kinds. Thank heaven you and I have our choices of what businesses we will attach ourselves to. If beekeeping sounds like your thing, go after it.

# Annotated Bibliography

Adams, John F., *Beekeeping: The Gentle Craft*. New York: Avon Books, 1972. This is an absorbing account about the nature and habits of bees, with additional information for the beginning beekeeper. It is written in a warm, personal style most interesting to read.

Aebi, Ormond and Harry, *The Art & Adventure of Beekeeping*. Santa Cruz, California: Unity Press, 1975. Father and son operate a very successful honey-producing business. This is their story of championship beekeeping. Easy and interesting reading.

Andersen, Arthur W., *Bee Prepared with Honey*. Bountiful, Utah: Horizon Publishers, 1975. Arthur Andersen was 92 years young when he wrote this book about his experiences with raising bees. He gives helpful information to beginning beekeepers. Recipes are included in the book.

Boy Scouts of America, *Beekeeping*. North Brunswick, N.J.: Boy Scouts of America, 1957, a 48-page pamphlet with limited information for beginning beekeeping. Written especially for the Boy Scout program.

Caron, Dewey M., *Beekeeping in Maryland*. College Park, Md.: Cooperative Extension Service, Bulletin 223, 1971. Although this 40-page pamphlet was prepared for beekeeping in Maryland, much of the information is helpful to beginning beekeepers everywhere.

Dadant, C. P., *First Lessons in Beekeeping*. Hamilton, Ill.: American Bee Journal, 1968. This 121-page booklet has much information for the novice about the habits of bees and beginning beekeeping.

Hanson, Louise G. and Davis, Lily A., *The Basic Beekeeping and Honey Book*. New York: David McKay Company, Inc., 1977. Lily and I wrote this book not only for the novice beekeeper but also for people who prefer to cook without sugar or any other sweetener except honey. It does not contain information about beekeeping as a profession.

Nye, W. P., *Nectar and Pollen Plants of Utah*. Logan, Utah: Utah State University Monograph Series, vol. 18, Number 3, March, 1971. Most helpful to any beekeeper in Utah, this 82-page pamphlet tells about the location of nectar and pollen plants in Utah.

Taylor, Richard, *The Joys of Beekeeping*. New York: St. Martin's Press, 1974. This 200-page book is aptly named. If you want to lose yourself in a well-written, happy book about beekeeping, this book is a must.

# Index